Undergraduate Texts in Mathematics

Springer

New York
Berlin
Heidelberg
Barcelona
Budapest
Hong Kong
London
Milan
Paris
Santa Clara
Singapore
Tokyo

Undergraduate Texts in Mathematics

Anglin: Mathematics: A Concise History and Philosophy.
Readings in Mathematics.

Anglin/Lambek: The Heritage of Thales.
Readings in Mathematics.

Apostol: Introduction to Analytic Number Theory. Second edition.

Armstrong: Basic Topology.

Armstrong: Groups and Symmetry.

Axler: Linear Algebra Done Right. Second edition.

Beardon: Limits: A New Approach to Real Analysis.

Bak/Newman: Complex Analysis. Second edition.

Banchoff/Wermer: Linear Algebra Through Geometry. Second edition.

Berberian: A First Course in Real Analysis.

Brémaud: An Introduction to Probabilistic Modeling.

Bressoud: Factorization and Primality Testing.

Bressoud: Second Year Calculus.
Readings in Mathematics.

Brickman: Mathematical Introduction to Linear Programming and Game Theory.

Browder: Mathematical Analysis: An Introduction.

Buskes/van Rooij: Topological Spaces: From Distance to Neighborhood.

Cederberg: A Course in Modern Geometries.

Childs: A Concrete Introduction to Higher Algebra. Second edition.

Chung: Elementary Probability Theory with Stochastic Processes. Third edition.

Cox/Little/O'Shea: Ideals, Varieties, and Algorithms. Second edition.

Croom: Basic Concepts of Algebraic Topology.

Curtis: Linear Algebra: An Introductory Approach. Fourth edition.

Devlin: The Joy of Sets: Fundamentals of Contemporary Set Theory. Second edition.

Dixmier: General Topology.

Driver: Why Math?

Ebbinghaus/Flum/Thomas: Mathematical Logic. Second edition.

Edgar: Measure, Topology, and Fractal Geometry.

Elaydi: Introduction to Difference Equations.

Exner: An Accompaniment to Higher Mathematics.

Fine/Rosenberger: The Fundamental Theory of Algebra.

Fischer: Intermediate Real Analysis.

Flanigan/Kazdan: Calculus Two: Linear and Nonlinear Functions. Second edition.

Fleming: Functions of Several Variables. Second edition.

Foulds: Combinatorial Optimization for Undergraduates.

Foulds: Optimization Techniques: An Introduction.

Franklin: Methods of Mathematical Economics.

Gordon: Discrete Probability.

Hairer/Wanner: Analysis by Its History.
Readings in Mathematics.

Halmos: Finite-Dimensional Vector Spaces. Second edition.

Halmos: Naive Set Theory.

Hämmerlin/Hoffmann: Numerical Mathematics.
Readings in Mathematics.

Hijab: Introduction to Calculus and Classical Analysis.

Hilton/Holton/Pedersen: Mathematical Reflections: In a Room with Many Mirrors.

Iooss/Joseph: Elementary Stability and Bifurcation Theory. Second edition.

Isaac: The Pleasures of Probability.
Readings in Mathematics.

Sterling K. Berberian

A First Course in Real Analysis

With 19 Illustrations

 Springer

Sterling K. Berberian
Department of Mathematics
The University of Texas at Austin
Austin, TX 78712
USA

On the cover: Formula for the fundamental theorem of calculus.

Mathematics Subject Classification (1991): 26-01

Library of Congress Cataloging-in-Publication Data
Berberian, Sterling K., 1926–
    A first course in real analysis/Sterling K. Berberian
      p. cm. — (Undergraduate texts in mathematics)
    Includes bibliographical references (p. –) and index.
    ISBN 0-387-94217-3 (New York). — ISBN 3-540-94217-3 (Berlin)
    1. Mathematical analysis. 2. Numbers, Real. I. Title.
    II. Series.
    QA300.B457   1994
    515 – dc20   93-46020

Printed on acid-free paper.

Production managed by Francine McNeill; manufacturing supervised by Genieve Shaw.
Photocomposed copy prepared using the author's TeX files.
Printed and bound by R.R. Donnelley and Sons, Harrisonburg, VA.
Printed in the United States of America.

9 8 7 6 5 4 3 2 (Corrected second printing, 1998)

ISBN 0-387-94217-3 Springer Verlag New York Berlin Heidelberg
ISBN 3-540-94217-3 Springer Verlag Berlin Heidelberg New York   SPIN 10662888

For Carol

Preface

Mathematics is the music of science, and real analysis is the Bach of mathematics. There are many other foolish things I could say about the subject of this book, but the foregoing will give the reader an idea of where my heart lies.

The present book was written to support a first course in real analysis, normally taken after a year of elementary calculus. Real analysis is, roughly speaking, the modern setting for Calculus, "real" alluding to the field of real numbers that underlies it all. At center stage are functions, defined and taking values in sets of real numbers or in sets (the plane, 3-space, etc.) readily derived from the real numbers; a first course in real analysis traditionally places the emphasis on real-valued functions defined on sets of real numbers.

The agenda for the course: (1) start with the axioms for the field of real numbers, (2) build, in one semester and with appropriate rigor, the foundations of calculus (including the "Fundamental Theorem"), and, along the way, (3) develop those skills and attitudes that enable us to continue learning mathematics on our own. Three decades of experience with the exercise have not diminished my astonishment that it can be done.

This paragraph is about prerequisites. The first three sections of the Appendix summarize the notations and terminology of sets and functions, and a few principles of elementary logic, that are the language in which the subject of the book is expressed. Ideally, the student will have met such matters in a preliminary course (on almost any subject) that provides some experience in the art of theorem-proving. Nonetheless, proofs are written out in sufficient detail that the book can also be used for a "first post-calculus, theorem-proving course"; a couple of days of orientation on material in the Appendix will be needed, and the going will be slower, at least at the beginning, but it can be done. Some topics will have to be sacrificed to accommodate the slower pace; I recommend sacrificing all of Chapters 10 (Infinite series) and 11 (Beyond the Riemann integral), the last four sections of Chapter 9, and Theorem 8.5.5 (used only in one of the sections of Chapter 9 just nominated for omission).

Chapter 11 is an attempt to bring some significant part of the theory of the Lebesgue integral—a powerful generalization of the integral of elementary calculus—within reach of an undergraduate course in analysis, while avoiding the sea of technical details that usually accompanies it. It will not be easy to squeeze this chapter into a one-semester course—perhaps it should not be tried (I never did in my classes)—but it is good to have a glimpse of what lies over the horizon. Probably the best use for this chapter is a self-study project for the student headed for a graduate course in real analysis; just reading the chapter introduction should be of some help in navigating the above-mentioned sea of details.

Each section is followed by an exercise set. The division of labor between text and exercises is straightforward: no proof in the text depends on the solution of an exercise and, throughout Chapters 1–10, the exercises are meant to be solvable using concepts and theorems already covered in the text (the very few exceptions to this rule are clearly signaled). On the other hand, many of the 'exercises' in Chapter 11 are included primarily for the reader's information and do require results not covered in the text; where this is the case, references to other sources are provided.

Some exercises contain extensions and generalizations (for example, the Riemann-Stieltjes integral and metric space topology) of material in the text proper; an exceptionally well-prepared class can easily integrate such topics into the mainstream. Other exercises offer alternative proofs of theorems proved in the text. Typically, exercises are for 'hands-on' practice, to strengthen the reader's grasp of the concepts introduced in that section; the easiest are good candidates for inclusion on a test, which is how many of them came into being.

Experience persuades me that sequential convergence is the most natural way to introduce limits. This form of the limit concept has much intuitive support (decimal expansions, the half-the-distance-to-the-wall story, etc.), but I acknowledge that the decision is subjective and that I have not always taught the course that way. The equivalent epsilon-delta and 'neighborhood' formulations of limits (superior to sequences in some situations) are also discussed fully.

My observation is that the introductory course in analysis, more than any other in the undergraduate mathematics curriculum, is a course that makes a difference. The right course at the right time, distilling millennia of human reflection into a brief semester's time, it is a wonderful learning machine whose output is, no less, Mathematicians.

Austin, Texas   Sterling Berberian
Summer, 1993

Contents

Contents

CHAPTER 1

Axioms for the Field $\mathbb{R}$ of Real Numbers

What are the real numbers and what can they do for us? The goal of this chapter is to give a working answer to the first question (the remaining chapters address the second). The present chapter should be regarded as a leisurely definition; read no more than is necessary to have a feeling for the axioms.

The *field of real numbers* is a set $\mathbb{R}$ of elements $a, b, c, \ldots$ (called real numbers), whose properties will be specified by a list of axioms. The axioms come in three groups: the *field axioms*, the *order axioms* and the *completeness axiom*.

The field axioms, entirely algebraic in character, have to do with properties of addition and multiplication. The order axioms pertain to relative 'size' (greater than, less than). The completeness axiom is an order axiom so important that it merits being split off from the others for emphasis.

1.1. The Field Axioms

1.1.1. *Definition.* A **field** is a set F such that, for every pair of elements a, b of F, there are defined elements $a + b$ and ab of F, called the *sum* and *product* of a and b, subject to the following axioms:

(A_1) $(a + b) + c = a + (b + c)$ (*associative law* for addition);

(A_2) $a + b = b + a$ (*commutative law* for addition);

(A_3) there is a unique element $0 \in F$ such that $a + 0 = a$ for all $a \in F$ (existence of a *zero element*);

(A_4) for each $a \in F$ there exists a unique element of F, denoted $-a$, such that $a + (-a) = 0$ (existence of *negatives*);

(M_1) $(ab)c = a(bc)$ (*associative law* for multiplication);

(M_2) $ab = ba$ (*commutative law* for multiplication);

1

(M$_3$) there is a unique element 1 of F, different from 0, such that $1a = a$ for all $a \in F$ (existence of a *unity element*);

(M$_4$) for each **nonzero** $a \in F$ there exists a unique element of F, denoted a^{-1}, such that $aa^{-1} = 1$ (existence of *reciprocals*);

(D) $a(b + c) = ab + ac$ (*distributive law*).

Properties A$_1$–A$_4$ pertain exclusively to addition, properties M$_1$–M$_4$ are their analogues for multiplication, and the distributive law is a joint property of addition and multiplication. It is instructive to notice that there are fields other than $\mathbb{R}$ (which has yet to be defined!).

1.1.2. *Example*. The field of *rational numbers* is the set $\mathbb{Q}$ of fractions,

$$\mathbb{Q} = \{m/n : m \text{ and } n \text{ integers}, n \neq 0\},$$

with the usual operations

$$\frac{m}{n} + \frac{m'}{n'} = \frac{mn' + nm'}{nn'},$$

$$\frac{m}{n} \cdot \frac{m'}{n'} = \frac{mm'}{nn'}.$$

The fraction $0/1$ serves as zero element, $1/1$ as unity element, $(-m)/n$ as the negative of m/n, and n/m as the reciprocal of m/n (assuming m and n both nonzero).

1.1.3. *Example*. The smallest field consists of two elements 0 and 1, where $1 + 1 = 0$ and all other sums and products are defined in the expected way (for example, $1 + 0 = 1$, $0 \cdot 1 = 0$).

1.1.4. *Example*. Let F be any field. Write $F[t]$ for the set of all polynomials $p(t) = a_0 + a_1 t + \ldots + a_n t^n$ in an indeterminate t, with coefficients a_k in F, and write $F(t)$ for the set of all 'fractions' $p(t)/q(t)$ with $p(t), q(t) \in F[t]$ and $q(t)$ not the zero polynomial. With sums and products defined by the same formulas as in 1.1.2, $F(t)$ is a field; it is called the field of *rational forms* over F.

1.1.5. *Example*. Write $\mathbb{Q} + i\mathbb{Q}$ for the set of all expressions $\alpha = r + is$ ($r, s \in \mathbb{Q}$). If $\alpha = r + is$ and $\alpha' = r' + is'$ are two such expressions, $\alpha = \alpha'$ means that $r = r'$ and $s = s'$. Sums and products are defined by the formulas

$$\alpha + \alpha' = (r + r') + i(s + s'),$$

$$\alpha\alpha' = (rr' - ss') + i(rs' + sr').$$

It is straightforward to verify that $\mathbb{Q} + i\mathbb{Q}$ is a field (called the field of *Gaussian rationals*), with $0 + i0$ serving as zero element, $-r + i(-s)$ as the negative of $r + is$, $1 + i0$ as unity element, and

$$\frac{r}{r^2 + s^2} + i\left(\frac{-s}{r^2 + s^2}\right)$$

as the reciprocal of $r + is$ (assuming at least one of r and s nonzero). Abbreviating $r + i0$ as r, we can regard $\mathbb{Q}$ as a subset of $\mathbb{Q} + i\mathbb{Q}$; abbreviating $0 + i1$ as i, we have $i^2 = -1$.

Here are some of the most frequently used properties of a field:

1.1.6. Theorem. *Let* F *be a field,* a *and* b *elements of* F.

(1) $a + a = a \Leftrightarrow a = 0$.
(2) $a0 = 0$ *for all* a.
(3) $-(-a) = a$ *for all* a.
(4) $a(-b) = -(ab) = (-a)b$ *for all* a *and* b.
(5) $(-a)^2 = a^2$ *for all* a.
(6) $ab = 0 \Rightarrow a = 0$ *or* $b = 0$; *in other words,*
(6') $a \neq 0$ & $b \neq 0 \Rightarrow ab \neq 0$.
(7) $a \neq 0$ & $b \neq 0 \Rightarrow (ab)^{-1} = a^{-1}b^{-1}$.
(8) $(-1)a = -a$ *for all* a.
(9) $-(a + b) = (-a) + (-b)$ *for all* a *and* b.
(10) *Defining* $a - b$ *to be* $a + (-b)$, *we have* $-(a - b) = b - a$.

Proof. (1) If $a + a = a$, add $-a$ to both sides:

$$(a + a) + (-a) = a + (-a)$$
$$a + (a + (-a)) = a + (-a) \quad [\text{axiom A}_1]$$
$$a + 0 = 0 \quad [\text{axiom A}_4]$$
$$a = 0. \quad [\text{axiom A}_3]$$

Thus $a + a = a \Rightarrow a = 0$, and the converse is immediate from axiom A_3.

(2) By axiom (D), $a0 = a(0 + 0) = a0 + a0$, so $a0 = 0$ by (1).

(3) $(-a) + a = a + (-a) = 0$, so $a = -(-a)$ by the uniqueness part of A_4.

(4) $0 = a0 = a[b + (-b)] = ab + a(-b)$, so $a(-b) = -(ab)$ by A_4; it follows that $(-a)b = b(-a) = -(ba) = -(ab)$.

(5) Citing (4) twice, we have $(-a)(-a) = -[a(-a)] = -[-(aa)] = aa$, in other words $(-a)^2 = a^2$. {Shorthand: $a^2 = aa$, $a^3 = a^2a$, $a^4 = a^3a$, etc.}

(6), (6'), (7) If a and b are nonzero, then $(ab)(a^{-1}b^{-1}) = (aa^{-1})(bb^{-1}) = 1 \cdot 1 = 1 \neq 0$; it follows from (2) that ab must be nonzero, and $(ab)^{-1} = a^{-1}b^{-1}$ follows from uniqueness in axiom M_4.

(8) $(-1)a = -(1a) = -a$.

(9) $-(a + b) = (-1)(a + b) = (-1)a + (-1)b = (-a) + (-b)$.

(10) $-(a - b) = -[a + (-b)] = (-a) + (-(-b)) = (-a) + b = b + (-a) = b - a$. $\Diamond$

1.1.7. *Notations.* In the rational field $\mathbb{Q}$, $m/n = m'/n'$ means that $mn' = nm'$; abbreviating $m/1$ as m, the set $\mathbb{Z}$ of integers can be regarded as a subset of $\mathbb{Q}$. For a nonzero integer n,

$$n(1/n) = (n/1)(1/n) = n/n = 1/1 = 1,$$

thus $1/n = n^{-1}$. The fractional notation is useful in an arbitrary field
F: one writes a/b for ab^{-1}, where $a, b \in$ F and $b \neq 0$. (This is the
multiplicative analogue of subtraction.)

Exercises

1. Let F and G be fields. A *monomorphism* of F into G is an
injective mapping $\varphi : \text{F} \to \text{G}$ that preserves sums and products: $\varphi(a+b)$
$= \varphi(a) + \varphi(b)$ and $\varphi(ab) = \varphi(a)\varphi(b)$ for all a, b in F. If, in addition, φ
is surjective (therefore bijective), it is called an *isomorphism* of F onto G
(and when F = G it is called an *automorphism* of F). Fields F and G
are said to be *isomorphic*, written F $\cong$ G, if there exists an isomorphism
of F onto G.

(i) If $\varphi : \text{F} \to \text{G}$ is an isomorphism of fields, then the inverse mapping
$\varphi^{-1} : \text{G} \to \text{F}$ is also an isomorphism. Thus, F $\cong$ G $\Rightarrow$ G $\cong$ F.

(ii) If $\varphi : \text{F} \to \text{G}$ is a monomorphism of fields, then $\varphi(1) = 1$. {Hint:
Apply φ to the equation $1^2 = 1$.}

(iii) The field $\mathbb{Q}$ of rational numbers is *not* isomorphic to the field $\mathbb{Q}(t)$
of rational forms over $\mathbb{Q}$. {Hint: Using (ii), argue that if $\varphi : \mathbb{Q} \to \mathbb{Q}(t)$ is
a monomorphism, then $\varphi(r) = r$ for all $r \in \mathbb{Q}$.}

2. In the field F = $\mathbb{Q} + i\mathbb{Q}$ of Gaussian rationals (1.1.5), if $\alpha = r + is$
define $\bar{\alpha} = r - is$ (called the *conjugate* of α). Show that the mapping
$\varphi : \text{F} \to \text{F}$ defined by $\varphi(\alpha) = \bar{\alpha}$ is an automorphism of F (in the sense of
Exercise 1) and that, for every nonzero element α, $\alpha^{-1} = (r^2 + s^2)^{-1}\bar{\alpha}$.

1.2. The Order Axioms

What sets ℝ apart from other fields are its order properties; in technical
terms, ℝ is a 'complete ordered field'. We look first at the concept of
'ordered field' (the question of completeness is taken up in §1.4).

1.2.1. *Definition.* An **ordered field** is a field F having a subset P of
nonzero elements, called *positive*, such that
(O$_1$) $a, b \in$ P $\Rightarrow$ $a + b \in$ P,
(O$_2$) $a, b \in$ P $\Rightarrow$ $ab \in$ P,
(O$_3$) $a \in$ F, $a \neq 0$ $\Rightarrow$ either $a \in$ P or $-a \in$ P, but not both.

In words, the sum and product of positive elements are positive; for each
nonzero element a, exactly one of a and $-a$ is positive.

For elements a, b of F, we write $a < b$ (or $b > a$) if $b - a \in$ P.

1.2.2. *Remarks.* With notations as in 1.2.1,

$$b \in \text{P} \iff b > 0,$$
$$-a \in \text{P} \iff a < 0.$$

Elements a with $a < 0$ are called *negative*. Properties O_1 and O_2 may be written

$$a > 0 \ \& \ b > 0 \ \Rightarrow \ a + b > 0 \ \& \ ab > 0.$$

Property O_3 yields the following: If $a, b \in F$, and $a \neq b$ (in other words, $a - b \neq 0$) then either $a > b$ or $a < b$ but not both. Thus, for any pair of elements a, b of F, *exactly one* of the following three statements is true:

$$a < b, \ a = b, \ a > b.$$

This form of O_3 is called the *law of trichotomy*.

The familiar properties of inequalities hold in any ordered field:

1.2.3. Theorem. *In an ordered field,*
  (1) $a < a$ *is impossible*;
  (2) *if* $a < b$ *and* $b < c$ *then* $a < c$;
  (3) $a < b \ \Leftrightarrow \ a + c < b + c$;
  (4) $a < b \ \Leftrightarrow \ -a > -b$;
  (5) $a < 0 \ \& \ b < 0 \ \Rightarrow \ ab > 0$;
  (6) $a < 0 \ \& \ b > 0 \ \Rightarrow \ ab < 0$;
  (7) $a < b \ \& \ c > 0 \ \Rightarrow \ ca < cb$;
  (8) $a < b \ \& \ c < 0 \ \Rightarrow \ ca > cb$;
  (9) $a \neq 0 \ \Rightarrow \ a^2 > 0$;
  (10) $1 > 0$;
  (11) $a + 1 > a$;
  (12) $a > 0 \ \Rightarrow \ a^{-1} > 0$.

Proof. (1) In the notations of 1.2.1, $a - a = 0 \notin P$, therefore $a < a$ cannot hold.
  (2) $c - a = (c - b) + (b - a)$ is the sum of two positive elements, therefore $a < c$.
  (3) $(b + c) - (a + c) = b - a$.
  (4) $-a - (-b) = b - a$.
  (5) $ab = (-a)(-b)$ is the product of two positive elements.
  (6) $0 - ab = (-a)b$ is the product of positives, therefore $ab < 0$.
  (7) $cb - ca = c(b - a)$.
  (8) $ca - cb = (-c)(b - a)$.
  (9) $a^2 = aa = (-a)(-a)$ is the product of two positives.
  (10) $1 = 1^2 > 0$ by (9).
  (11) $(a + 1) - a = 1 > 0$.
  (12) If $a > 0$ then $aa^{-1} = 1 > 0$ precludes $a^{-1} < 0$ by (6). ◊

1.2.4. *Definition.* In any ordered field, one defines $2 = 1 + 1$, $3 = 2 + 1$, $4 = 3 + 1$, etc.
  We have $0 < 1 < 2 < 3 < 4 < \ldots$ by 1.2.3.

1.2.5. *Definition.* In an ordered field, we write $a \leq b$ (also $b \geq a$) if either $a < b$ or $a = b$. An element a such that $a \geq 0$ is said to be *nonnegative.*

The expected properties of this notation are easily verified. {For example, $a \leq b$ & $b \leq c \Rightarrow a \leq c$; $a < b$ & $c \geq 0 \Rightarrow ca \leq cb$, etc.}

1.2.6. *Example.* The rational field $\mathbb{Q}$ is ordered, with

$$P = \{m/n : m \text{ and } n \text{ positive integers}\}$$

as the set of positive elements. One sees from 1.2.3 and 1.2.4 that $\mathbb{Q}$ is the 'smallest' ordered field.

1.2.7. *Example.* The field of Gaussian rationals (1.1.5) is not orderable, because $i^2 = -1$. {In an ordered field, nonzero squares are positive and -1 is negative, so -1 can't be a square.}

Near the surface, and very useful:

1.2.8. **Theorem.** *Let* F *be an ordered field,* a *and* b **nonnegative** *elements of* F, n *any positive integer.*
  (i) $a < b \Leftrightarrow a^n < b^n$;
  (ii) $a = b \Leftrightarrow a^n = b^n$;
  (iii) $a > b \Leftrightarrow a^n > b^n$.

Proof. (i), $\Rightarrow$: By assumption, $0 \leq a < b$; we have to prove that $a^n < b^n$ for every positive integer n. The proof is by induction on n, the case $n = 1$ being the given inequality. Assuming $a^k < b^k$, consider the identity

$$b^{k+1} - a^{k+1} = b(b^k - a^k) + (b - a)a^k;$$

the right side is positive because $b > 0$, $b^k - a^k > 0$, $b - a > 0$ and $a^k \geq 0$, thus $a^{k+1} < b^{k+1}$.

The implication "$\Rightarrow$" of (iii) follows on interchanging the roles of a and b, and the implication "$\Rightarrow$" of (ii) is obvious. The implication "$\Rightarrow$" is thus valid for each of the three statements.

The implications "$\Leftarrow$" come free of charge by trichotomy! For example, suppose $a^n < b^n$; the assertion is that $a < b$. The alternatives, $a = b$ or $a > b$, are unacceptable since they would imply either $a^n = b^n$ or $a^n > b^n$ (both of which are false). $\Diamond$

Exercises

1. In an ordered field, $0 < a < b \Rightarrow b^{-1} < a^{-1}$. {Hint: $a^{-1} - b^{-1} = a^{-1}(b - a)b^{-1}$.}

2. In an ordered field, $a^2 + b^2 = 0 \Rightarrow a = b = 0$. Similarly for a sum of n squares ($n = 3, 4, 5, \ldots$). {Hint: $a^2 + b^2 \geq a^2$.}

3. In an ordered field, if $a, b, c \geq 0$ and $a \leq b + c$, then

$$\frac{a}{1+a} \leq \frac{b}{1+b} + \frac{c}{1+c}.$$

4. Write out the proof of 1.2.8 in full.

5. Let F be an ordered field, a and b any elements of F (not necessarily positive), and n an *odd* positive integer. Prove:
(i) $a < b \Leftrightarrow a^n < b^n$;
(ii) $a = b \Leftrightarrow a^n = b^n$;
(iii) $a > b \Leftrightarrow a^n > b^n$.
{Hint: The heart of the matter is to prove "$\Rightarrow$" of (i). Consider first the case that $a < 0$ (and $b < 0$ or $b = 0$ or $b > 0$).}

6. Let $\mathbb{Q}(t)$ be the field of rational forms over $\mathbb{Q}$ (1.1.4) and let S be the set of all nonzero elements $r(t) = p(t)/q(t)$ of $\mathbb{Q}(t)$ such that the leading coefficients of the polynomials p and q have the same sign, that is, such that $p(t) = a_0 + a_1 t + \ldots + a_m t^m$, $q(t) = b_0 + b_1 t + \ldots + b_n t^n$ with $a_m b_n > 0$. {Equivalently, r can be written in the form $r = p/q$, where p and q are polynomials with integral coefficients and the leading coefficients are both positive.}
(i) $\mathbb{Q}(t)$ is an ordered field, with S as set of positive elements.
(ii) For the ordering defined in (i), $n1 < t$ for every positive integer n.

7. In an ordered field, if $a < b$ and if $x = (a+b)/2 = (a+b)2^{-1}$ then $a < x < b$. (So to speak, the average of two elements lies between them.)

8. In an ordered field, describe the set of elements x such that $(2x + 1)(3x - 5) > 0$.

9. In an ordered field, if $a < b$ then $a < \frac{1}{3}a + \frac{2}{3}b < b$; more generally,

$$a < ra + (1 - r)b < b$$

whenever $a < b$ and $0 < r < 1$.

10. If a, b are elements of an ordered field, such that $a \leq b + c$ for every $c > 0$, then $a \leq b$. {Hint: Assume to the contrary and cite Exercise 7.}

11. In an ordered field, $ab \leq [\frac{1}{2}(a + b)]^2$ for all a, b.

12. In an ordered field, $a^2 - ab + b^2 \geq 0$ for all a, b. {Hint: Divide $a + b$ into $a^3 + b^3$, then consider separately the cases $ab \leq 0$, $ab > 0$.}

13. (Bernoulli's inequality) In an ordered field, let x be any element such that $x \geq -1$. Prove that

$$(1 + x)^n \geq 1 + nx$$

for every positive integer n. If $x > -1$ and $x \neq 0$, then $(1+x)^n > 1 + nx$ for all $n \geq 2$. {Hint: Induction.}

14. In an ordered field, $(1+x)^{2n} \geq 1 + 2nx$ for every element x and every positive integer n. {Hint: Exercise 13.}

15. Prove that in an ordered field,
(i) $x \geq 0 \Rightarrow (1+x)^n \geq 1 + nx + \frac{1}{2}n(n-1)x^2$ for every positive integer n.

(ii) $0 \leq x \leq 1 \Rightarrow (1-x)^n \leq 1 - nx + \frac{1}{2}n(n-1)x^2$ for every positive integer n.

(iii) If $a_i \geq 0$ for $i = 1, \ldots, n$ then

$$\prod_{i=1}^{n}(1+a_i) \geq 1 + \sum_{i=1}^{n} a_i .$$

16. In an ordered field, if $x \geq 0$ and n is an integer ≥ 2, then

$$(1+x)^n \geq 1 + nx + \frac{1}{4}n^2 x^2 > \frac{1}{4}n^2 x^2 .$$

17. If n is an integer that is not the square of an integer, then n is not the square of a rational number. {An easy special case: n a prime number. Still easier: the case that $n = 2$.}

18. In an ordered field, let P be the set of elements > 0. Prove that P has neither a smallest element nor a largest element.

19. In an ordered field, $n < 2^n$ for every positive integer n. {Hint: Induction.}

20. In an ordered field,

$$\left(1 - \frac{1}{n^2}\right)^n > 1 - \frac{1}{n}$$

for every integer $n \geq 2$ (Exercise 13); infer (by factoring the left side) that

$$\left(1 + \frac{1}{n-1}\right)^{n-1} < \left(1 + \frac{1}{n}\right)^n .$$

21. In an ordered field, suppose $a_1 \leq a_2 \leq a_3 \leq \ldots$. Let $b_n = (a_1 + \ldots + a_n)/n$ be the 'average' of $a_1, \ldots, a_n$. Prove that $b_1 \leq b_2 \leq b_3 \leq \ldots$.

1.3. Bounded Sets, LUB and GLB

1.3.1. *Definition.* Let F be an ordered field. A nonempty subset A of F is said to be

(i) **bounded above** if there exists an element $K \in F$ such that $x \leq K$ for all $x \in A$ (such an element K is called an *upper bound* for A);

(ii) **bounded below** if there exists an element $k \in F$ such that $k \leq x$ for all $x \in A$ (such an element k is called a *lower bound* for A);

(iii) **bounded** if it is both bounded above and bounded below;

(iv) **unbounded** if it is not bounded.

1.3.2. *Examples.* Let F be an ordered field, a and b elements of F with $a \leq b$. Each of the following subsets of F is bounded, with a serving as a lower bound and b as an upper bound:

$$[a, b] = \{x \in F : a \leq x \leq b\}$$
$$(a, b) = \{x \in F : a < x < b\}$$
$$[a, b) = \{x \in F : a \leq x < b\}$$
$$(a, b] = \{x \in F : a < x \leq b\}.$$

Such subsets of F are called *intervals*, with *endpoints* a and b; more precisely, $[a, b]$ is called a *closed interval* (because it contains the endpoints) and (a, b) is called an *open interval* (because it doesn't); the intervals $[a, b)$ and $(a, b]$ are called 'semiclosed' (or 'semi-open'). {Caution: If $F = \mathbb{Q}$ then the term 'interval' loses some of its intuitive meaning (an interval in $\mathbb{Q}$ is considerably more ventilated—all those missing irrationals!—than the familiar intervals on the real line); from the next section onward, the term is used only in the context of the field $F = \mathbb{R}$ of real numbers (cf. §4.1).}

Note that $(a, a) = [a, a) = (a, a] = \varnothing$ because $a < a$ is impossible. On the other hand, $[a, a] = \{a\}$ (the set whose only element is a).

1.3.3. *Example.* An ordered field is neither bounded above nor bounded below (for example, any proposed upper bound K is topped by $K + 1$).

1.3.4. *Example.* In an ordered field F, the interval $[0, 1]$ has a largest element but $[0, 1)$ does not; for, if a is any element of $[0, 1)$ then $x = (a + 1)/2$ is a larger element of $[0, 1)$ (§1.2, Exercise 7).

Note that 1 is an upper bound for $[0, 1)$, but nothing smaller will do: if $a < 1$ then $[0, 1)$ contains an element x larger than a (if $a < 0$ let $x = 1/2$; if $0 \leq a < 1$ let $x = (a + 1)/2$). This prompts the following definition:

1.3.5. *Definition.* Let F be an ordered field, A a nonempty subset of F. We say that A *has a least upper bound* in F if there exists an element $M \in F$ such that

(a) M is an upper bound for A, that is, $x \leq M$ for all $x \in A$;

(b) nothing smaller than M is an upper bound for A; that is,

$$M' < M \implies \exists x \in A \ni x > M',$$

or, contrapositively,

$$M' \text{ an upper bound for A} \;\Rightarrow\; M \le M'.$$

Since anything larger than an upper bound is also an upper bound, conditions (a) and (b) can be combined into a single condition:

$$M' \text{ is an upper bound for } A \;\Leftrightarrow\; M' \ge M.$$

If such a number M exists, it is obviously unique (it is the smallest element of the set of upper bounds for A); it is called the **least upper bound** (or **supremum**) of A, written

$$M = \mathrm{LUB}\,A, \text{ or } M = \sup A.$$

1.3.6. *Remarks.* As observed in 1.3.4, in any ordered field $\sup [0,1) = 1$; note that the supremum of a set need not belong to the set. A set that is bounded above need not have a least upper bound (Exercise 3).

There is an analogous concept for sets that are bounded below:

1.3.7. *Definition.* Let A be a nonempty subset of an ordered field F. We say that A *has a greatest lower bound* in F if there exists an element $m \in F$ such that
(a) m is a lower bound for A,
(b) if m' is a lower bound for A then $m \ge m'$.
Such an element m is unique; it is called the **greatest lower bound** (or **infimum**) of A, written

$$m = \mathrm{GLB}\,A, \text{ or } m = \inf A.$$

There is a natural 'duality' between sups and infs:

1.3.8. Theorem. *Let* F *be an ordered field,* A *a nonempty subset of* F; *write*
$$-A = \{-x:\ x \in A\}$$
for the set of negatives of the elements of A. *Let* $c \in F$. *Then:*
(i) c *is an upper bound for* A $\Leftrightarrow$ $-c$ *is a lower bound for* $-A$;
(ii) c *is a lower bound for* A $\Leftrightarrow$ $-c$ *is an upper bound for* $-A$;
(iii) *If* A *has a least upper bound, then* $-A$ *has a greatest lower bound and* $\inf(-A) = -(\sup A)$.
(iv) *If* A *has a greatest lower bound, then* $-A$ *has a least upper bound and* $\sup(-A) = -(\inf A)$.

Proof. (i), (ii) The mapping $x \mapsto -x$ is a bijection $F \to F$ that reverses order:
$$a < b \;\Leftrightarrow\; -a > -b.$$

The condition
$$x \le c \text{ for all } x \in A$$

is therefore equivalent to the condition

$$-c \le y \text{ for all } y \in -A.$$

This proves (i), and the proof of (ii) is similar.

(iii) Suppose A has a least upper bound α; we know from (i) that $-\alpha$ is a lower bound for $-A$ and we have to show that it is larger than all others. Let k be any lower bound for $-A$. For all $a \in A$ we have $k \le -a$, so $a \le -k$; this shows that $-k$ is an upper bound for A, therefore $\alpha \le -k$, so $-\alpha \ge k$ as we wished to show.

The proof of (iv) is similar to that of (iii). ◊

Exercises

1. Let F be an ordered field, $a \in F$. The subset

$$\{x \in F : x < a\}$$

is bounded above, but not below; the subset

$$\{x \in F : x > a\}$$

is bounded below, but not above. Each of the intervals of 1.3.2 is the intersection of two unbounded sets; for example,

$$[a, b) = \{x : x \ge a\} \cap \{x : x < b\}.$$

2. In the field $\mathbb{Q}$ of rational numbers, the subset $\mathbb{P} = \{1, 2, 3, \ldots\}$ is not bounded above.
{Hint: If $K = m/n$ with m and n positive integers, then $m+1 > K$; every candidate K for upper bound is topped by a positive integer.}

3. In the ordered field $\mathbb{Q}(t)$ of §1.2, Exercise 6, the subset $\mathbb{P} = \{1, 2, 3, \ldots\}$ is bounded; it has no least upper bound.
{Hint: If $r = p/q \in \mathbb{Q}(t)$ is any upper bound for $\mathbb{P}$, then $2n < r$ for all $n \in \mathbb{P}$, so $r/2$ is an upper bound for $\mathbb{P}$ smaller than r.}

4. If, in an ordered field, $m = \inf A$ and $m' > m$, show that there exists an element $a \in A$ such that $m \le a < m'$ (cf. 1.3.5).

1.4. The Completeness Axiom (Existence of LUB's)

The property that distinguishes the field of real numbers from all other ordered fields is an assumption about the existence of least upper bounds:

1.4.1. Definition. An ordered field is said to be **complete** if it satisfies the following condition: *Every nonempty subset that is bounded above has a least upper bound.*

The crucial questions: Do such fields exist? If so, how many? The point of departure of the present course is the assumption that the answers are "yes" and "one".[1] More precisely, we assume that *there exists a complete ordered field* $\mathbb{R}$ *and that it is unique in the sense that every complete ordered field is isomorphic to* $\mathbb{R}$ (cf. §1.1, Exercise 1). For convenient reference:

1.4.2. Definition. $\mathbb{R}$ is a complete ordered field; its elements are called **real numbers.**

In other words, $\mathbb{R}$ is a set with two operations (addition and multiplication) satisfying the *field axioms* (1.1.1), the *order axioms* (1.2.1) and the *completeness axiom* (1.4.1).

1.4.3. Examples. The ordered field $\mathbb{Q}(t)$ is *not* complete (§1.3, Exercise 3). The rational field $\mathbb{Q}$ isn't complete either, but it takes a little more work to prove it (2.2.4).

1.4.4. Remarks. With the field $\mathbb{R}$ of real numbers defined axiomatically, we shouldn't take the integers and rational numbers for granted. We're going to anyway, to save time; thus the statements that follow are definitely *not* self-evident.

The set of **positive integers** is the set

$$\mathbb{P} = \{1, 2, 3, \ldots\},$$

where $2 = 1 + 1$, $3 = 2 + 1$, etc. {The "etc." and the three dots "..." hide all the difficulties! A formal definition of $\mathbb{P}$ is given in Exercise 6; for the moment, think of $\mathbb{P}$ informally as the set of 'finite sums' whose terms are all equal to 1.} Every positive integer is > 0, and 1 is the smallest. The set $\mathbb{P}$ is closed under addition and multiplication.

The set of **natural numbers** is the set

$$\mathbb{N} = \{0\} \cup \mathbb{P} = \{0, 1, 2, 3, \ldots\};$$

$\mathbb{N}$ is closed under addition and multiplication.

---

[1]It may seem surprising that something so fundamental as a consensus on the treatment of the real numbers is to some extent a matter of faith and/or hope—hope that the discovery of a hidden inconsistency does not bring the whole structure tumbling down. Given enough assumptions about sets and rules for manipulating them, generations of mathematicians have been able to persuade themselves that it is possible to construct a complete ordered field. (The hard part is existence; once this is settled, the proof of uniqueness is relatively straightforward.) The pioneer who did it first was the German mathematician Richard Dedekind (1831-1916). There have always been dissenters of genius to the general consensus, beginning with Dedekind's fellow countryman Leopold Kronecker (1823-1891).

The set of **integers** is the set $\mathbb{Z}$ of all differences of positive integers,

$$\mathbb{Z} = \{m - n : \ m, n \in \mathbb{P}\} \, ;$$

$\mathbb{Z}$ is closed under the operations $x + y$, xy and $-x$. The set of positive elements of $\mathbb{Z}$ is precisely the set $\mathbb{P}$, therefore $\mathbb{Z} = \{0\} \cup \mathbb{P} \cup (-\mathbb{P})$, where $-\mathbb{P} = \{-n : \ n \in \mathbb{P}\}$.

The set of **rational numbers** is the set

$$\mathbb{Q} = \{m/n : \ m, n \in \mathbb{Z}, \ n \neq 0\} \, ,$$

where $m/n = mn^{-1}$; $\mathbb{Q}$ contains sums, products, negatives and reciprocals (of its nonzero elements), thus $\mathbb{Q}$ is itself a field (a 'subfield' of $\mathbb{R}$).

We have the inclusions $\mathbb{P} \subset \mathbb{N} \subset \mathbb{Z} \subset \mathbb{Q} \subset \mathbb{R}$.

Besides the overt axioms for $\mathbb{R}$ (field, order, completeness) we propose to accept a somewhat vague hidden one, namely, that $\mathbb{P}$ is 'equal' to the set of 'ordinary positive integers' with which we have been friends for a long time. There is essentially one path to righteousness: we should (1) set down axioms for the set of positive integers (for example, *Peano's axioms*),[2] (2) show that there is essentially only one such set, (3) give an unambiguous definition of the set $\mathbb{P}$ defined informally above, and (4) verify that $\mathbb{P}$ satisfies the axioms in question. As long as we are assuming $\mathbb{R}$ to be given axiomatically, an acceptable shortcut is to take only steps (3) and (4) of the path; this is carried out in the Appendix (§A.4).

For a first reading, a time-saving, if not totally virtuous, alternative (I recommend it!) is to provisionally accept the informal description of $\mathbb{P}$, glance at the 'plausibility arguments' in Exercises 3–5 for the key properties of $\mathbb{P}$ (well-ordering property, principle of mathematical induction), and move on to Chapter 2.[3]

Exercises

1. Let A and B be nonempty subsets of $\mathbb{R}$ that are bounded above, and write

$$A + B = \{a + b : \ a \in A, \ b \in B\} \, .$$

Prove that $A + B$ is bounded above and that $\sup(A + B) = \sup A + \sup B$.

---

[2]Cf. E. Landau, *Foundations of analysis* [Chelsea, New York, 1951].

[3]In the next chapter (2.2.4) we use the fact from elementary number theory that 2 is not the square of a rational number (cf. §1.2, Exercise 17) to prove the existence of irrational numbers (real numbers that are not rational), but the more spectacular proof sketched in §2.6, Exercise 1 requires no number theory at all.

{Hint: Let $\alpha = \sup A$, $\beta = \sup B$ and observe that $\alpha + \beta$ is an upper bound for $A + B$. It suffices to show that if c is any upper bound for $A + B$ then $\alpha + \beta \leq c$. If $a \in A$ then $a + b \leq c$ for all $b \in B$, so $c - a$ is an upper bound for B; it follows that $\beta \leq c - a$, thus $a \leq c - \beta$ for all $a \in A$.}

2. Let A and B be nonempty sets of *positive* real numbers that are bounded above, and write

$$AB = \{ab : a \in A, \, b \in B\}.$$

Prove that AB is bounded above and that $\sup(AB) = (\sup A)(\sup B)$. {Hint: Imitate the proof of Exercise 1, using division instead of subtraction.}

3. Argue (informally!) that the set of positive elements of $\mathbb{Z}$ is $\mathbb{P}$.

{Hint: Suppose $m, n \in \mathbb{P}$ and $m - n > 0$, say $m - n = x \in \mathbb{R}$, $x > 0$. Think of each of m, n as a sum of 1's. The sum for m must have more terms than the sum for n (otherwise, cancellation of the 1's on the left side of $m = n + x$ would exhibit 0 as a sum of positives); thus $m - n$ is a sum of 1's, hence is an element of $\mathbb{P}$.}

4. Argue (informally!) that every nonempty subset S of $\mathbb{P}$ has a smallest element (*well-ordering property* of $\mathbb{P}$).

{Hint: Assume to the contrary that S contains a decreasing sequence of elements $n_1 > n_2 > n_3 > \dots$. Since $n_1 - n_2 > 0$ and $n_1 - n_2$ is an integer, $n_1 - n_2 \geq 1$ (cf. Exercise 3). Similarly $n_2 - n_3 \geq 1$, so $n_1 - n_3 \geq 2$. 'Continuing', $n_1 - n_{k+1} \geq k$, so $k < n_1$ for all $k \in \mathbb{P}$, contrary to §1.3, Exercise 2.}

5. (*Principle of mathematical induction*) If T is a subset of $\mathbb{P}$ such that $1 \in T$ and such that $n \in T \Rightarrow n + 1 \in T$, then $T = \mathbb{P}$.

{Hint: If $T \neq \mathbb{P}$, the smallest element of $S = \mathbb{P} - T$ is an embarrassment.}

6. Here is a formal definition of the set $\mathbb{P}$ described informally in 1.4.4. There exist sets $S \subset \mathbb{R}$ such that (a) $1 \in S$, and (b) $x \in S \Rightarrow x + 1 \in S$ (for example, $\mathbb{R}$ has these properties). *Define* $\mathbb{P}$ to be the *intersection* of all such subsets S of $\mathbb{R}$. For $n \in \mathbb{P}$ write $n' = n + 1$. Prove: (i) $1 \in \mathbb{P}$; (ii) $n \in \mathbb{P} \Rightarrow n' \in \mathbb{P}$; (iii) if $1 \in S \subset \mathbb{P}$ and $n \in S \Rightarrow n' \in S$, then $S = \mathbb{P}$; (iv) if $m, n \in \mathbb{P}$ and $m' = n'$ then $m = n$; (v) for all $n \in \mathbb{P}$, $1 \neq n'$. {Hint for (v): Using (iii), show that $n > 0$ for all $n \in \mathbb{P}$.}

In other words, the set $\mathbb{P}$, with distinguished element 1 and distinguished mapping $n \mapsto n'$, satisfies *Peano's axioms* for the positive integers.

CHAPTER 2

First Properties of $\mathbb{R}$

Chapter 1 lays the foundation; what are we building? Our most ambitious objective is to give a logically rigorous presentation of Calculus, learning as much as possible along the way. Everything has to come from the axioms for the field of real numbers $\mathbb{R}$ (that in fact it does seems a miracle!). The properties of $\mathbb{R}$ developed in this chapter lie very close to the axioms; they require some ingenuity but no essential new concepts (such as limits).

2.1. Dual of the Completeness Axiom (Existence of GLB's)

The existence of least upper bounds is an assumption (completeness axiom); the existence of greatest lower bounds then comes free of charge:

2.1.1. **Theorem.** *If* A *is a nonempty subset of* $\mathbb{R}$ *that is bounded below, then* A *has a greatest lower bound, namely,* $\inf A = -\sup(-A)$.

Proof. The set $-A = \{-a : a \in A\}$ is nonempty and bounded above (1.3.8, (ii)), so it has a least upper bound by the completeness axiom. By (iii) of 1.3.8, $-(-A) = A$ has a greatest lower bound and $\inf A = -\sup(-A)$. $\Diamond$

A valuable application:

2.1.2. **Corollary.** $\inf \{1/n : n \in \mathbb{P}\} = 0$.

15

Proof. Let $A = \{1/n : n \in \mathbb{P}\}$. By (12) of 1.2.3, A is bounded below by 0, so A has a greatest lower bound a (2.1.1) and $0 \le a$. On the other hand, $a \le 1/2n$ for all positive integers n, so $2a$ is also a lower bound for A; it follows that $2a \le a$, therefore $a \le 0$ and finally $a = 0$. $\diamond$

Exercises

1. If A and B are nonempty subsets of $\mathbb{R}$ such that $a \le b$ for all $a \in A$ and $b \in B$, then A is bounded above, B is bounded below, and $\sup A \le \inf B$.

2. If A is a nonempty subset of $\mathbb{R}$ that is bounded above, then $\sup A = -\inf(-A)$.

3. If $A \subset \mathbb{R}$ is nonempty and bounded above, and if B is the set of all upper bounds for A, show that $\sup A = \inf B$, then state and prove the 'dual' result.

4. If a is a positive real number then $\inf\{a/n : n \in \mathbb{P}\} = 0$. {Hint: Modify the proof of 2.1.2.}

5. Find $\inf\{(-1)^n + 1/n : n \in \mathbb{P}\}$.

6. Let $\mathcal{S}$ be any set of intervals I of $\mathbb{R}$ (defined as in 1.3.2) and let $J = \bigcap \mathcal{S}$ be their intersection, that is,

$$ J = \bigcap \{I : I \in \mathcal{S}\} = \{x \in \mathbb{R} : x \in I \text{ for all } I \in \mathcal{S}\}. $$

Prove that there exist real numbers α and β, with $\alpha \le \beta$, such that

$$ (\alpha, \beta) \subset J \subset [\alpha, \beta] ; $$

in particular, J is an interval (possibly empty, possibly degenerate).
{Hint: Assuming J nonempty, let $x \in J$. Let A be the set of all left endpoints of intervals $I \in \mathcal{S}$, B the set of right endpoints. If $I \in \mathcal{S}$ has left endpoint a and right endpoint b, then $(a, b) \subset I \subset [a, b]$; infer that $a \le x \le b$, so x is an upper bound for A and a lower bound for B. The numbers $\alpha = \sup A$, $\beta = \inf B$ meet the requirements.}

2.2. Archimedean Property

In every ordered field, $1 < 2 < 3 < \ldots$, therefore $1 > 1/2 > 1/3 > \ldots$ (see 1.2.4 and §1.2, Exercise 1); for every $y > 0$, we thus have $y > y/2 > y/3 > \ldots$. We are habituated to expecting the elements y/n $(n = 1, 2, 3, \ldots)$ to be 'arbitrarily small' in the sense that, for every $x > 0$, there is an n for which y/n is smaller than x. In actuality, there exist ordered fields in which it can happen that $y/n \geq x > 0$ for all n, that is, the elements y/n $(n = 1, 2, 3, \ldots)$ are 'buffered away from 0' by the element x (cf. §1.2, Exercise 6). The property at the heart of such considerations is the following:

2.2.1. *Definition.* An ordered field is said to be **Archimedean** if, for each pair of elements x, y with $x > 0$, there exists a positive integer n such that $nx > y$. (Think of x as a 'unit of measurement'. Each element y can be surpassed by a sufficiently large multiple of the unit of measurement.)

2.2.2. *Theorem.* *The field $\mathbb{R}$ of real numbers is Archimedean.*

Proof. Let x and y be real numbers, with $x > 0$. If $y \leq 0$ then $1x > y$. Assuming $y > 0$, we seek a positive integer n such that $1/n < x/y$; the alternative is that $0 < x/y \leq 1/n$ for every positive integer n, and this is contrary to $\inf\{1/n : n \in \mathbb{P}\} = 0$ (2.1.2). $\Diamond$

2.2.3. *Example.* The field $\mathbb{Q}(t)$ of rational forms over $\mathbb{Q}$, ordered as in §1.2, Exercise 6, is *not* Archimedean.

The message of 2.2.2 is that, in an ordered field, completeness implies the Archimedean property. The converse is false:

2.2.4. *Theorem.* *The field $\mathbb{Q}$ of rational numbers is Archimedean but not complete.*

Proof. The Archimedean property for $\mathbb{Q}$ is an immediate consequence of 2.2.2 (see also Exercise 1). We have to exhibit a nonempty subset A of $\mathbb{Q}$ that is bounded above but has no least upper bound in $\mathbb{Q}$; the core of the proof is the fact that 2 is not the square of a rational number (cf. §1.2, Exercise 17).

Let $A = \{r \in \mathbb{Q} : r > 0 \text{ and } r^2 < 2\}$; for example $1 \in A$, so A is nonempty. If $r \in \mathbb{Q}$ and $r \geq 2$ then $r^2 \geq 4 > 2$, so $r \notin A$; stated contrapositively, $r < 2$ for all $r \in A$, thus A is bounded above.

We now show that A *has no largest element.* The strategy is simple: given any element r of A, we shall produce a larger element of A. For this, it suffices to find a positive integer n such that $r + 1/n \in A$, that is, $(r + 1/n)^2 < 2$; by elementary algebra, this is equivalent to the condition

$$(*) \qquad\qquad n(2 - r^2) > 2r + 1/n.$$

Since $2 - r^2 > 0$, the Archimedean property yields a positive integer n such that $n(2 - r^2) > 2r + 1$; but $2r + 1 \geq 2r + 1/n$, so the condition $(*)$ is verified.

There are positive elements r of $\mathbb{Q}$ such that $r^2 > 2$ (for example $r = 2$); we show next that *there is no smallest such element r*. Given any $r \in \mathbb{Q}$ with $r > 0$ and $r^2 > 2$, we shall produce a positive element of $\mathbb{Q}$

that is smaller than r but whose square is also larger than 2. It suffices to find a positive integer n such that $r - 1/n > 0$ and $(r - 1/n)^2 > 2$, equivalently,

$$(**) \qquad nr > 1 \quad \text{and} \quad n(r^2 - 2) > 2r - 1/n.$$

Since $r > 0$ and $r^2 - 2 > 0$, the Archimedean property yields a positive integer n such that *both*

$$nr > 1 \quad \text{and} \quad n(r^2 - 2) > 2r$$

(choose an n for each inequality, then take the larger of the two); but $2r > 2r - 1/n$, so the condition (**) is verified.

Finally, we assert that A has no least upper bound in $\mathbb{Q}$. Assume to the contrary that A has a least upper bound t in $\mathbb{Q}$. We know that $t^2 \neq 2$ (2 is not the square of a rational number) and $t > 0$ (because $1 \in A$). Let us show that each of the possibilities $t^2 < 2$ and $t^2 > 2$ leads to a contradiction.

If $t^2 < 2$ then $t \in A$; but then t would be the largest element of A, contrary to our earlier observation that no such element exists.

If $t^2 > 2$ then, as observed above, there exists a rational number s such that $0 < s < t$ and $s^2 > 2$. Since t is supposedly the *least* upper bound of A and s is smaller than t, s can't be an upper bound for A. This means that there exists an element r of A with $s < r$; but then $s^2 < r^2 < 2$, contrary to $s^2 > 2$. $\Diamond$

Exercises

1. Give an elementary proof of the Archimedean property for $\mathbb{Q}$ (that is, a proof that doesn't depend on the existence of $\mathbb{R}$).

{Hint: The essential case is $x, y \in \mathbb{Q}$ with $x > 0$ and $y > 0$. Write $x = a/b$ and $y = c/d$ with a, b, c, d positive integers. The inequality $nx > y$ is equivalent to $nad > bc$. Try $n = 1 + bc$.}

2. Show that for every real number $y > 0$,

$$\bigcap_{n=1}^{\infty} (0, y/n] = \varnothing.$$

{Hint: An element x belonging to the left side would have to satisfy the inequalities $0 < x \leq y/n$ for every positive integer n.}

3. Show that the subset $\mathbb{P} = \{1, 2, 3, \ldots\}$ of $\mathbb{R}$ is not bounded above.

2.3. Bracket Function

A useful application of the Archimedean property is that every real number can be sandwiched between a pair of successive integers:

2.3.1. **Theorem.** *For each real number* x, *there exists a unique integer* n *such that* $n \leq x < n+1$.

Proof. Uniqueness: The claim is that a real number x can't belong to the interval $[n, n+1)$ for two distinct values of n. If m and n are distinct integers, say $m < n$, then $n - m$ is an integer and is > 0, therefore $n - m \geq 1$ (cf. 1.4.4); thus $m + 1 \leq n$ and it follows that the intervals $[m, m+1)$ and $[n, n+1)$ can have no element x in common.

Existence: Let $x \in \mathbb{R}$. By the Archimedean property, there exists a positive integer j such that $j1 > -x$, that is, $j + x > 0$. It will suffice to find an integer k such that $j + x \in [k, k+1)$, for this would imply that $x \in [k - j, k - j + 1)$; changing notation, we can suppose that $x > 0$.

Let $S = \{ k \in \mathbb{P} : k1 > x \}$. By the Archimedean property S is nonempty, so S has a smallest element m by the 'well-ordering principle' (every nonempty set of positive integers has a smallest element). Since $m \in S$, we have $m > x$.

If $m = 1$ then $0 < x < 1$ and the assertion is proved with $n = 0$.

If $m > 1$ then $m - 1$ is a positive integer smaller than m, so it can't belong to S; this means that $m - 1 \leq x$, thus $x \in [m - 1, m)$ and $n = m - 1$ fills the bill. ◊

2.3.2. **Definition.** With notations as in the theorem, the integer n is denoted $[x]$ and the function $\mathbb{R} \to \mathbb{Z}$ defined by $x \mapsto [x]$ is called the **bracket function** (or the *greatest integer function*, since $[x]$ is the largest integer that is $\leq x$).

Exercises

1. Calculate $[2.3]$, $[-2.3]$ and $[-2]$.

2. Sketch the graph of each of the following functions $f : \mathbb{R} \to \mathbb{R}$.
(i) $f(x) = [x]$. {Hint: 'Step-function'.}
(ii) $f(x) = [2x]$.
(iii) $f(x) = [-x + 3]$.

3. If $x = .234$ then $[100x + .5]/100 = .23$; if $x = .235$ then $[100x + .5]/100 = .24$. Infer a general principle about 'rounding off'.

2.4. Density of the Rationals

Between any two reals, there's a rational:

2.4.1. Theorem. *If x and y are real numbers such that $x < y$, then there exists a rational number r such that $x < r < y$.*

Proof. Since $y - x > 0$, by the Archimedean property there exists a positive integer n such that $n(y - x) > 1$, that is, $1/n < y - x$. Think of $1/n$ as a 'unit of measurement', small enough for the task at hand; we propose to find a multiple of $1/n$ that lands between x and y.

By 2.3.1 (applied to the real number nx), there is an integer m such that $m \leq nx < m + 1$; then $m/n \leq x$ and

$$x < (m+1)/n = m/n + 1/n \leq x + 1/n < x + (y - x) = y,$$

so $r = (m+1)/n$ meets the requirements of the theorem. $\Diamond$

The conclusion of the theorem is expressed by saying that the rational field $\mathbb{Q}$ is *everywhere dense* in $\mathbb{R}$. So to speak, there are 'lots' of rational numbers. A sobering thought: Are there any real numbers that *aren't* rational? The answer is yes and the proof is implicit in Theorem 2.2.4: the set A described there is nonempty and bounded above, so it has a least upper bound u in $\mathbb{R}$; if u were rational, then it would be a least upper bound for A in the ordered field $\mathbb{Q}$, contrary to what was shown in 2.2.4.

2.4.2. *Definition.* A real number that is not rational is called an **irrational number**; thus, the irrational numbers are the elements of the difference set

$$\mathbb{R} - \mathbb{Q} = \{x \in \mathbb{R} : x \notin \mathbb{Q}\}.$$

Exercises

1. If x is any real number, prove that

$$x = \sup\{r \in \mathbb{Q} : r < x\} = \inf\{s \in \mathbb{Q} : x < s\}.$$

2. (i) True or false (explain): If x is rational and y is irrational, then $x + y$ is irrational. {Hint: $y = (x + y) - x$.}

(ii) True or false (explain): If x and y are both irrational, then so is $x + y$.

(iii) True or false (explain): If x is irrational and y is a nonzero rational, then xy is irrational.

3. Show that if x and y are real numbers with $x < y$, then there exists an irrational number t such that $x < t < y$.

{Hint: If z is irrational, then so are $-z$ and z/n ($n = 1, 2, 3, \ldots$). Adapt the proof of 2.4.1.}

4. (i) Give an example of a set of rational numbers whose supremum is irrational.

(ii) Give an example of a set of irrational numbers whose supremum is rational.

2.5. Monotone Sequences

The concept of sequence is an old friend from elementary calculus (the sequence of partial sums of an infinite series, the sequence of derivatives of a function, etc.). The familiar examples are sequences of numbers and sequences of functions, but, in principle, the elements that make up a sequence can be drawn from any nonempty set:

2.5.1. Definition. If X is a set and if, for each positive integer n, an element x_n of X is given, we say that we have a **sequence** of elements of X, or 'a sequence *in* X', whose **nth term** is x_n.

Various notations are used to indicate sequences, for example

$$(x_n), \ (x_n)_{n \in \mathbb{P}}, \ (x_n)_{n \geq 1}, \ (x_n)_{n=1,2,3,\ldots}$$

(nothing sacred about the letter x, of course!).

Informally, a sequence of elements of a set is an unending list x_1, x_2, $x_3, \ldots$ of elements (not necessarily distinct) of the set. More formally, we can think of it as a function $f : \mathbb{P} \to X$, where we have chosen to write x_n instead of $f(n)$ for the element of X corresponding to the positive integer n. Another notation that stresses the functional aspect of a sequence: $n \mapsto x_n \ (n \in \mathbb{P})$.

In the notation (x_n), the integers n are called the *indices*. Sometimes index sets other than $\mathbb{P}$ are appropriate; for example, $(a_n)_{n \in \mathbb{N}}$ for the coefficients of a power series $\sum_{n=0}^{\infty} a_n x^n$.[1]

In this section, the focus is on certain special sequences of real numbers:

2.5.2. Definition. A sequence (a_n) in $\mathbb{R}$ is said to be **increasing** if $a_1 \leq a_2 \leq a_3 \leq \ldots$, that is, if $a_n \leq a_{n+1}$ for all $n \in \mathbb{P}$; **strictly increasing** if $a_n < a_{n+1}$ for all n; **decreasing** if $a_1 \geq a_2 \geq a_3 \geq \ldots$; and **strictly decreasing** if $a_n > a_{n+1}$ for all n.

A sequence that is either increasing or decreasing is said to be **monotone**; more precisely, one speaks of sequences that are 'monotone increasing' or 'monotone decreasing'.

---

[1] The ultimate generalization: For any function $f : I \to X$ of a set I into a set X, we are free to write x_i (or a_i, etc.) instead of $f(i)$, to call I a set of indices, and to use $(x_i)_{i \in I}$ as an alternative notation for the function f; we then speak of $(x_i)_{i \in I}$ as a *family* of elements of X, *indexed* by the set I. The sequences defined in 2.5.1 are the families indexed by $\mathbb{P}$.

If (a_n) is an increasing sequence, we write $a_n \uparrow$, and if it is a decreasing sequence we write $a_n \downarrow$. (No special notation is offered for 'strictly monotone' sequences.)

So far, this is just noodling with notations; here is a concept with some substance:

2.5.3. Definition. If (a_n) is an increasing sequence in $\mathbb{R}$ such that the set $A = \{a_n : n \in \mathbb{P}\}$ is *bounded above*, and if $a = \sup A$, then we write $a_n \uparrow a$; similarly, $a_n \downarrow a$ means that (a_n) is a decreasing sequence, the set $A = \{a_n : n \in \mathbb{P}\}$ is bounded below, and $a = \inf A$.

2.5.4. Example. $1/n \downarrow 0$. For, the sequence $(1/n)$ is decreasing (§1.2, Exercise 1) and $\inf\{1/n : n \in \mathbb{P}\} = 0$ (2.1.2).

2.5.5. Example. If $0 < c < 1$ then the sequence of powers (c^n) is strictly decreasing and $c^n \downarrow 0$.

{Proof: To see that (c^n) is strictly decreasing, multiply $0 < c < 1$ by c to get $0 < c^2 < c$. Do it again to get $0 < c^3 < c^2$, and so on. Let $a = \inf\{c^n : n \in \mathbb{P}\}$. We know that $a \geq 0$ and $c^n \downarrow a$; the problem is to show that $a = 0$. Since $a/c \leq c^n$ for all n (because $a \leq c^{n+1}$), it follows that $a/c \leq a$; thus $a(1 - c) \leq 0$, therefore $a \leq 0$ (because $1 - c > 0$) and finally $a = 0$.}

The notations of 2.5.3 have some useful formal properties:

2.5.6. Theorem. *If $a_n \uparrow a$ and $b_n \uparrow b$, then*
(i) $a_n + b_n \uparrow a + b$,
(ii) $-a_n \downarrow -a$,
(iii) $a_n + c \uparrow a + c$ *for every real number* c.

Proof. (i) It is clear that $(a_n + b_n)$ is an increasing sequence that is bounded above by $a + b$; the problem is to show that $a + b$ is the *least* upper bound.

Suppose $a_n + b_n \leq c$ for all n; we have to show that $a + b \leq c$, that is, $a \leq c - b$. Given any index m, it is enough to show that $a_m \leq c - b$, that is, $b \leq c - a_m$; thus, given any index n, we need only show that $b_n \leq c - a_m$, that is, $a_m + b_n \leq c$. Indeed, if p is the larger of m and n then $a_m + b_n \leq a_p + b_p \leq c$ by the assumed monotonicity.

(ii) This is immediate from (iii) of 1.3.8.
(iii) This is a special case of (i), with $b_n = c$ for all n. $\Diamond$

Exercises

1. If $a_n \downarrow a$ and $b_n \downarrow b$ then $a_n + b_n \downarrow a + b$.

2. If $a_n \uparrow a$, $b_n \uparrow b$ and $a_n \geq 0$, $b_n \geq 0$ for all n, then $a_n b_n \uparrow ab$.

3. If (d_n) is a sequence in the set $\{0, 1, 2, \ldots, 9\}$, interpret the 'decimal' $.d_1 d_2 d_3 \ldots$ as a real number in the interval $[0, 1]$.

{Hint: Consider $a_n = .d_1 d_2 \ldots d_n = d_1 d_2 \ldots d_n / 10^n$, where $d_1 d_2 \ldots d_n$ is the integer whose digits are $d_1, d_2, \ldots, d_n$.}

4. Suppose $a_n \uparrow a$ and $b_n \uparrow b$. Let $c = \max\{a, b\}$ be the larger of a and b, $c_n = \max\{a_n, b_n\}$ the larger of a_n and b_n. Prove that $c_n \uparrow c$.

5. Suppose $a_n \uparrow a$ and $b_n \uparrow b$. If $a_n \leq b_n$ for all n then $a \leq b$; is the converse true?

6. If $a_n \downarrow 0$ and $b > 0$ then $a_n b \downarrow 0$.

7. If $A = \{(-1)^n + 1/n : n \in \mathbb{P}\}$, find $\inf A$ and $\sup A$. {Hint: If $A = B \cup C$ then $\inf A = \inf\{\inf B, \inf C\}$.}

8. If $a_n \downarrow a$, $b_n \downarrow b$ and $a \geq 0$, $b \geq 0$, then $a_n b_n \downarrow ab$.

9. (i) If r is a real number such that $0 \leq r \leq 1$, then there exists a unique real number s such that $0 \leq s \leq 1$ and $s^2 = r$.
(ii) Infer from (i) that if $a \in \mathbb{R}$, $a \geq 0$, then there exists a unique $b \in \mathbb{R}$, $b \geq 0$, such that $b^2 = a$.
{Hints: (i) Writing $y = 1 - r$, $x = 1 - s$, we have $0 \leq y \leq 1$ and the problem is to find a real number x, $0 \leq x \leq 1$, such that $(1-x)^2 = 1-y$, that is, $x = \frac{1}{2}(y + x^2)$. The formulas $x_1 = 0$, $x_{n+1} = \frac{1}{2}[y + (x_n)^2]$ define recursively an increasing sequence (x_n) such that $0 \leq x_n \leq 1$. (This style of proof is called the "method of successive approximations".) (ii) Choose a positive integer m such that $m \geq a$ and consider $0 \leq a/m^2 \leq 1$.}

2.6. Theorem on Nested Intervals

A sequence of intervals (I_n) of $\mathbb{R}$ is said to be *nested* if $I_1 \supset I_2 \supset I_3 \supset \ldots$. As the intervals 'shrink' with increasing n, there is no assurance that there is any point that belongs to every I_n: consider, for example, $I_n = (0, 1/n]$ (§2.2, Exercise 2). However, if the intervals are *closed*, we can be sure that there's at least one survivor:

2.6.1. Theorem. *If (I_n) is a nested sequence of closed intervals, then the intersection of the I_n is nonempty. More precisely, if $I_n = [a_n, b_n]$, where $a_n \leq b_n$ and $I_1 \supset I_2 \supset I_3 \supset \ldots$, and if*

$$a = \sup\{a_n : n \in \mathbb{P}\}, \quad b = \inf\{b_n : n \in \mathbb{P}\},$$

then $a \leq b$ and $\bigcap_{n=1}^{\infty}[a_n, b_n] = [a, b]$.

Proof. {The notation $\bigcap_{n=1}^{\infty}[a_n, b_n]$ means the intersection $\bigcap \mathcal{S}$ of the set $\mathcal{S}$ of all the intervals $[a_n, b_n]$ (cf. the Appendix, A.2.9).}
From $[a_{n+1}, b_{n+1}] \subset [a_n, b_n]$ we see that

$$a_n \leq a_{n+1} \leq b_{n+1} \leq b_n;$$

it follows that the sequence (a_n) is increasing and bounded above (for example by b_1), whereas (b_n) is decreasing and bounded below (for example by a_1). If a and b are defined as in the statement of the theorem, we have $a_n \uparrow a$ and $b_n \downarrow b$ (2.5.3). By 2.5.6 (and its 'dual', with arrows reversed) we have $-b_n \uparrow -b$, so $a_n + (-b_n) \uparrow a + (-b)$, therefore $b_n - a_n \downarrow b - a$. Since $b_n - a_n \geq 0$ for all n, it follows that $b - a \geq 0$; then $a_n \leq a \leq b \leq b_n$, so $[a, b] \subset [a_n, b_n]$ for all n, therefore

$$[a, b] \subset \bigcap_{n=1}^{\infty} [a_n, b_n].$$

Conversely, if x belongs to every $[a_n, b_n]$ then $a_n \leq x \leq b_n$ for all n, therefore $a \leq x \leq b$ and we have shown that

$$\bigcap_{n=1}^{\infty} [a_n, b_n] \subset [a, b]. \diamondsuit$$

The following corollary is known as the *Theorem on nested intervals*:

2.6.2. **Corollary.** *If, in addition to the assumptions of the theorem,*

$$\inf(b_n - a_n) = 0,$$

then there exists exactly one point common to the intervals $[a_n, b_n]$, *that is,*

$$\bigcap_{n=1}^{\infty} [a_n, b_n] = \{c\}$$

for a suitable point c.

Proof. As shown in the proof of the theorem, $b_n - a_n \downarrow b - a$; but $b_n - a_n \downarrow 0$ by assumption, so $b = a$ and

$$\bigcap_{n=1}^{\infty} [a_n, b_n] = [a, a] = \{a\}. \diamondsuit$$

A surprising corollary is a nonconstructive[1] proof of the existence of irrational numbers, quite different from the proof in §2.4; the argument is sketched in Exercise 1.

---

[1]The argument persuades us that an irrational number exists, but gives us no clear idea as to its size.

Exercises

1. (i) The set of all rational numbers can be listed in a sequence $r_1, r_2, r_3, \ldots$ (in other words, there exists a surjection $\mathbb{P} \to \mathbb{Q}$).

{Hint: First list the positive rational numbers in a sequence $s_1, s_2, s_3, \ldots$, for example,

$$\frac{1}{1}, \frac{1}{2}, \frac{2}{1}, \frac{1}{3}, \frac{2}{2}, \frac{3}{1}, \ldots$$

(the fractions m/n with $m+n = 2$, then $m+n = 3$, then $m+n = 4$ and so on); the list $0, s_1, -s_1, s_2, -s_2, s_3, -s_3, \ldots$ then contains every rational number.}

(ii) There exists a nested sequence of closed intervals $[a_n, b_n]$ such that for every n, $r_n \notin [a_n, b_n]$.

{Hint: Call $[0, \frac{1}{3}]$ the *left third* of the interval $[0, 1]$, and $[\frac{2}{3}, 1]$ the *right third*, and similarly for any closed interval $[a, b]$. With notations as in (i), the point r_1 can't belong to both the left third and the right third of $[0, 1]$; let I_1 be a third that excludes it. Let I_2 be a third of I_1 that excludes r_2. Let I_3 be a third of I_2 that excludes r_3, and so on.}

(iii) With notations as in (ii), necessarily

$$\bigcap_{n=1}^{\infty} [a_n, b_n] = \{a\}$$

with a irrational. {Hint: The intersection is a closed interval $[a, b]$ (2.6.1) that excludes every rational number, so $a < b$ is impossible (2.4.1).}

2. If $x = .2847$ then

$$x \in [.284, .285] \subset [.28, .29] \subset [.2, .3].$$

Interpret a 'decimal' $.d_1 d_2 d_3 \ldots$ as the intersection of a nested sequence of closed intervals.

3. Let (a_n, b_n) be a nested sequence of open intervals, where $a_n < b_n$ for all n, and let $A = \bigcap (a_n, b_n)$ be their intersection.

(i) It can happen that $A = \varnothing$. (Example?)

(ii) If (a_n) is strictly increasing and (b_n) is strictly decreasing, then A is a *closed* interval (in particular, $A \neq \varnothing$). {Hint: Apply 2.6.1 to the closed intervals $[a_n, b_n]$.}

4. Let $[a, b]$ be a closed interval in $\mathbb{R}$, $a < b$, and let (x_n) be any sequence in $\mathbb{R}$. Prove that $[a, b]$ contains a real number not equal to any term of the sequence. {Hint: Cf. Exercise 1.}

2.7. Dedekind Cut Property

2.7.1. *Definition.* A **cut** (or *Dedekind cut*) of the real field ℝ is a pair (A, B) of nonempty subsets of ℝ such that every real number belongs to either A or B and such that $a < b$ for all $a \in A$ and $b \in B$. In symbols,

$$A \neq \varnothing, \quad B \neq \varnothing, \quad \mathbb{R} = A \cup B, \quad a < b \ (\forall a \in A, b \in B).$$

(It follows from the latter property that $A \cap B = \varnothing$.)

2.7.2. *Examples.* If $\gamma \in \mathbb{R}$ and

(*) $A = \{x \in \mathbb{R} : \ x \leq \gamma\}, \quad B = \{x \in \mathbb{R} : \ x > \gamma\}$

then (A, B) is a cut of ℝ (note that A has a largest element but B has no smallest); the pair

(**) $A = \{x \in \mathbb{R} : \ x < \gamma\}, \quad B = \{x \in \mathbb{R} : \ x \geq \gamma\}$

also defines a cut of ℝ (for which B has a smallest element but A has no largest).

The key fact about cuts of ℝ is that there are no other examples:

2.7.3. **Theorem.** *If* (A, B) *is a cut of* ℝ *then there exists a unique real number* γ *such that either the formulas* (*) *or the formulas* (**) *are verified.*

Proof. Uniqueness. The number γ is uniquely determined by the property of being either the largest element of A or the smallest element of B, according as we are in case (*) or in case (**).

Existence. Note that A is bounded above (by any element of B) and B is bounded below (by any element of A). Let

$$\alpha = \sup A, \quad \beta = \inf B.$$

If $a \in A$ then $a < b$ for all $b \in B$, therefore $a \leq \beta$; since $a \in A$ is arbitrary, $\alpha \leq \beta$. In fact $\alpha = \beta$, for if $\alpha < \beta$ then any number in the gap between α and β is an embarrassment: it is too large to belong to A, too small to belong to B. Write γ for the common value of α and β; by assumption, γ must belong to either A or B.

case 1: $\gamma \in A$.

Let's show that the formulas (*) are verified. At any rate,

$$A \subset \{x \in \mathbb{R} : \ x \leq \gamma\}, \quad B \subset \{x \in \mathbb{R} : \ x > \gamma\};$$

for, the first inclusion follows from $\gamma = \sup A$, and the second inclusion follows from $\gamma = \inf B$ and the fact that $\gamma \in B$ is ruled out by $\gamma \in A$.

These are 'self-improving' inclusions: they imply that both are in fact equalities. For example, if $x \leq \gamma$ then necessarily $x \in A$; the alternative is $x \in B$, unacceptable because it would imply $x > \gamma$. (The crux of the matter is that the left members of the inclusions have union $\mathbb{R}$, whereas the right members are disjoint.)

case 2: $\gamma \in B$.

One argues similarly that the formulas (**) are valid. {Alternatively, apply case 1 to the cut $(-B, -A)$ of $\mathbb{R}$, where $-B$ is the set of negatives of the elements of B, and similarly for $-A$.} $\Diamond$

The way cuts were originally introduced, and why Dedekind's name is associated with them, is explained in the exercises.

Exercises

1. Let A and B be nonempty sets of rational numbers, with $A \cup B = \mathbb{Q}$, such that $a < b$ for all $a \in A$, $b \in B$. Prove:

(i) There exists a unique real number γ such that either

(*) $$A = \{r \in \mathbb{Q} : r \leq \gamma\}, \quad B = \{r \in \mathbb{Q} : r > \gamma\}$$

or

(**) $$A = \{r \in \mathbb{Q} : r < \gamma\}, \quad B = \{r \in \mathbb{Q} : r \geq \gamma\}.$$

{Hint: Imitate the proof of 2.7.3.}

(ii) If γ is rational, then exactly one of the conditions (*) or (**) holds.

(iii) If γ is irrational, then both of the conditions (*) and (**) hold, and they may both be expressed as

$$A = \{r \in \mathbb{Q} : r < \gamma\}, \quad B = \{r \in \mathbb{Q} : r > \gamma\}.$$

2. With notations as in Exercise 1, the pair (A, B) is called a (Dedekind) *cut* of the rational field $\mathbb{Q}$.

(i) Show that the formulas

$$A = \{r \in \mathbb{Q} : r \leq 0\} \cup \{r \in \mathbb{Q} : r > 0 \text{ and } r^2 < 2\}$$
$$B = \{r \in \mathbb{Q} : r > 0 \text{ and } r^2 > 2\}$$

define a cut (A, B) of the rationals. {Hint: There is no rational number r with $r^2 = 2$.}

(ii) Let γ be the real number provided by Exercise 1. Show that $\gamma^2 = 2$. {Hint: As shown in the proof of 2.2.4, there is no largest positive number whose square is < 2, no smallest whose square is > 2.}

Dedekind's stroke of genius[1]: *Define* $\sqrt{2}$ *to be* this cut (A, B) of the rationals. There are still many loose ends to be looked after (defining sums and products of cuts, regarding cuts (*) and (**) that lead to the same real number as being 'equal', etc.), but these are largely housekeeping chores. The decisive stroke is the first: something new ($\sqrt{2}$) has been defined in terms of something old (a strategically chosen pair of subsets of $\mathbb{Q}$).

2.8. Square Roots

2.8.1. Theorem. *Every positive real number has a unique positive square root. That is, if $c \in \mathbb{R}$, $c > 0$, then there exists a unique $x \in \mathbb{R}$, $x > 0$, such that $x^2 = c$.*

Proof. Uniqueness: If x and y are positive real numbers such that $x^2 = c = y^2$ then $x = y$ by 1.2.8. {Explicitly, $0 = x^2 - y^2 = (x+y)(x-y)$ and $x + y > 0$, therefore $x - y = 0$.}

Existence: Given $c \in \mathbb{R}$, $c > 0$, the strategy is to construct a cut (A, B) of $\mathbb{R}$ for which the γ of 2.7.3 satisfies $\gamma^2 = c$. Let

$$A = \{x \in \mathbb{R} : x \leq 0\} \cup \{x \in \mathbb{R} : x > 0 \text{ and } x^2 < c\},$$
$$B = \{x \in \mathbb{R} : x > 0 \text{ and } x^2 \geq c\}.$$

Then $A \neq \varnothing$, $B \neq \varnothing$ (for example, $c + 1 \in B$) and $A \cup B = \mathbb{R}$ (obvious). Moreover, if $a \in A$ and $b \in B$ then $a < b$: this is trivial if $a \leq 0$, whereas if $a > 0$ then $a^2 < c \leq b^2$ implies $a < b$ (1.2.8). In summary, (A, B) is a cut of $\mathbb{R}$; let γ be the real number that defines the cut (2.7.3).

Note that A contains numbers > 0; for, if $c \geq 1$ then $1/2 \in A$ (because $1/4 < 1 \leq c$), whereas if $0 < c < 1$ then $c \in A$ (because $c^2 < c$). It follows that $\gamma > 0$.

Next, we assert that $\gamma \in B$; by the arguments in the preceding section, we need only show that A has no largest element. Assuming $a \in A$, let's find a larger element of A. If $a \leq 0$ then any positive element of A will do. Suppose $a > 0$. We know that $a^2 < c$; it will suffice to find a positive integer n such that $(a + 1/n)^2 < c$. The existence of such an n is shown by the argument in 2.2.4 (with 2 replaced by c).

From 2.7.3, we now know that

$$A = \{x \in \mathbb{R} : x < \gamma\}, \quad B = \{x \in \mathbb{R} : x \geq \gamma\}.$$

Since $\gamma \in B$ we have $\gamma^2 \geq c$; it remains only to show that $\gamma^2 \leq c$, that is, $\gamma^2 - c \leq 0$.

---

[1]R. Dedekind, *The theory of numbers* (translated from the German original) [Open Court Publ. Co., LaSalle, 1901; reprinted by Dover Publ. Co., New York].

By the Archimedean property, choose a positive integer N such that $N\gamma > 1$. For every integer $n \geq N$, we have $1/n \leq 1/N < \gamma$, so $\gamma - 1/n > 0$; since, moreover, $\gamma - 1/n$ belongs to A (it is $< \gamma$), it follows that $(\gamma - 1/n)^2 < c$, therefore

$$\gamma^2 - c < 2\gamma/n - 1/n^2 < 2\gamma/n.$$

Thus $(\gamma^2 - c)/2\gamma < 1/n$ for all $n \geq N$, and all the more for $1 \leq n < N$, consequently

$$(\gamma^2 - c)/2\gamma \leq \inf\{1/n : n \in \mathbb{P}\} = 0$$

(2.1.2); since $2\gamma > 0$ we conclude that $\gamma^2 - c \leq 0$, as we wished to show. $\diamondsuit$

2.8.2. *Definition.* With notations as in the theorem, x is called the **square root** of c and is denoted $\sqrt{c}$;[1] the definition is rounded out by defining $\sqrt{0} = 0$.

It follows that every nonnegative real number has a unique nonnegative square root.

Exercises

1. Let $A = \{x \in \mathbb{R} : x \geq 0\}$. Prove that the mapping $f : A \to A$ defined by $f(x) = x^2$ is bijective.

2. Infer the existence of irrational numbers from 2.8.1 (cf. §2.6, Exercise 1).

3. If $0 < a < b$ in $\mathbb{R}$, then $\sqrt{a} < \sqrt{b}$.

4. Prove that $\sqrt{2} + \sqrt{3}$ is irrational. Then prove that $\sqrt{n} + \sqrt{n+1}$ is irrational for every positive integer n. {Hint: §1.2, Exercise 17.}

5. Define a sequence (a_n) in $\mathbb{R}$ recursively as follows: $a_1 = \sqrt{2}$, $a_{n+1} = \sqrt{2 + a_n}$.
   (i) Show that (a_n) is strictly increasing. {Hint: Induction.}
   (ii) Show that $a_n < 2$ for every positive integer n.
   (iii) Let $a = \sup a_n$. Prove that $a = 2$. {Hint: Argue that $a^2 = a + 2$ (cf. §2.5, Exercise 2).}

6. If $\varphi : \mathbb{R} \to \mathbb{R}$ is a monomorphism of fields (§1.1, Exercise 1) then φ must be the identity mapping: $\varphi(x) = x$ for all $x \in \mathbb{R}$. Sketch of proof:
   (i) $\varphi(r) = r$ for every rational number r.
   (ii) $x \leq y \Rightarrow \varphi(x) \leq \varphi(y)$. {Hint: Write $y - x = z^2$ and apply φ.}

---

[1] One also writes $c^{1/2}$ for $\sqrt{c}$, a notation consistent with the general definition of powers in §6.2 and 9.5.13.

(iii) $\varphi(x) = x$ for every $x \in \mathbb{R}$. {Hint: If r and s are rationals with $r < x < s$, then $r \leq \varphi(x) \leq s$ by (i) and (ii); infer from §2.4, Exercise 1 that $x \leq \varphi(x) \leq x$.}

7. If $0 < x < y$ then $\sqrt{y} - \sqrt{x} < \sqrt{y - x}$.

2.9. Absolute Value

2.9.1. *Definition.* The **absolute value** of a real number a is the non-negative real number $|a|$ defined as follows:

$$|a| = \begin{cases} a & \text{if } a > 0 \\ 0 & \text{if } a = 0 \\ -a & \text{if } a < 0 \end{cases}$$

in other words,

$$|a| = \begin{cases} a & \text{if } a \geq 0 \\ -a & \text{if } a \leq 0. \end{cases}$$

Absolute value can be defined in any ordered field (1.2.1); its elementary properties follow at once from the properties of order:

2.9.2. Theorem. *For real numbers* a, b, c, x:
(1) $|a| \geq 0$.
(2) $|a|^2 = a^2$.
(3) *Properties* (1) *and* (2) *characterize* $|a|$: *if* $x \geq 0$ *and* $x^2 = a^2$ *then* $x = |a|$.
(4) $|a| = 0 \Leftrightarrow a = 0$; $|a| > 0 \Leftrightarrow a \neq 0$.
(5) $|a| = |b| \Leftrightarrow a^2 = b^2 \Leftrightarrow a = \pm b$.
(6) $|-a| = |a|$.
(7) $|ab| = |a| \, |b|$.
(8) $-|a| \leq a \leq |a|$; *more generally,*
(9) $|x| \leq c \Leftrightarrow -c \leq x \leq c$.
(10) $|a + b| \leq |a| + |b|$.
(11) $||a| - |b|| \leq |a - b|$.

Proof. (1), (2) and (4) are obvious from the definition of absolute value.
(3) If $x \geq 0$ and $x^2 = a^2$, that is, $x^2 = |a|^2$, then $x = |a|$ by 1.2.8.
(5) and (6) follow easily from (1)–(3).
(7) If $x = |a| \, |b|$, then $x^2 = |a|^2 |b|^2 = a^2 b^2 = (ab)^2$, therefore $x = |ab|$ by (3).
(8) If $a \geq 0$ then $-|a| = -a \leq 0 \leq a = |a|$, whereas if $a \leq 0$ then $-|a| = -(-a) = a \leq 0 \leq |a|$.
(9) If $-c \leq x \leq c$ then both $-x \leq c$ and $x \leq c$; but $|x|$ is either x or $-x$, so $|x| \leq c$. Conversely, if $|x| \leq c$ then $-c \leq -|x| \leq x \leq |x| \leq c$.

(10) Addition of the inequalities

$$-|a| \le a \le |a|, \quad -|b| \le b \le |b|$$

yields $-(|a| + |b|) \le a + b \le |a| + |b|$, so $|a + b| \le |a| + |b|$ by (9).

(11) Let $x = |a| - |b|$. Then $|a| = |(a - b) + b| \le |a - b| + |b|$, thus $x \le |a - b|$. Interchanging a and b, we have $-x \le |b - a| = |a - b|$, therefore $|x| \le |a - b|$. ◊

When the points of a line are labeled with real numbers in the usual way (with the 'origin' labeled 0), $|a|$ may be interpreted as the distance from the origin to the point labeled a. For example, $|\pm 5| = 5$ means that either of the points labeled 5 and -5 has distance 5 from the origin (Figure 1).

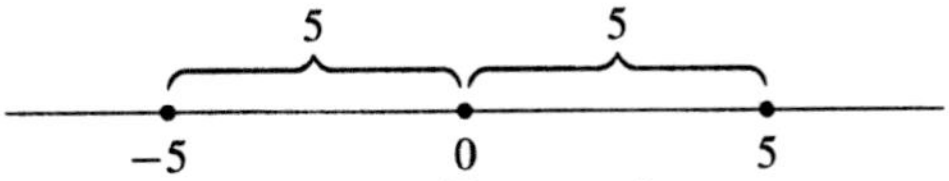

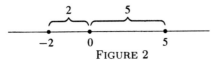

FIGURE 1

More generally, the distance between any two real numbers can be defined in terms of absolute value:

2.9.3. *Definition.* For real numbers a, b the **distance** from a to b is defined to be $|a - b|$. One also writes $d(a, b) = |a - b|$; the function $d : \mathbb{R} \times \mathbb{R} \to \mathbb{R}$ defined by this formula is called the (usual) *distance function* on $\mathbb{R}$.

For example (Figure 2), if $a = -2$ and $b = 5$, then $|a-b| = |-2-5| = 7$.

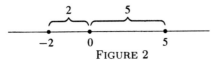

FIGURE 2

As we see in the next chapter, distance—equivalently, absolute value—is at the bottom of convergence (which is at the bottom of everything else!).[1]

Exercises

1. For any real numbers a and b,

$$\sup\{a, b\} = \frac{1}{2}(a + b + |a - b|),$$

$$\inf\{a, b\} = \frac{1}{2}(a + b - |a - b|),$$

---

[1] We could easily dispense with the concept of distance, but the geometric language it suggests is useful. For example, instead of saying that $|a - b|$ is 'small' we have the option of saying that a is 'near' b; instead of '$|a - b|$ becomes arbitrarily small' we can say that 'a approaches b', etc.

where $\sup\{a,b\} = \max\{a,b\}$ is the larger of a and b, and $\inf\{a,b\} = \min\{a,b\}$ is the smaller.

2. Let A be a nonempty set of real numbers. Prove that A is bounded in the sense of 1.3.1 if and only if there exists a positive real number K such that $|x| \le K$ for all $x \in A$. {Hint: If $A \subset [a,b]$ try $K = \max\{|a|,|b|\}$.}

3. Show that $|a+b| = |a| + |b|$ if and only if $ab \ge 0$. {Hint: After squaring, the equation is equivalent to $ab = |ab|$.} Note that if a and b are nonzero, then $ab > 0$ means that a and b have the same sign.

4. With notations as in 2.9.3, verify the following properties of the distance function d:
  (i) $d(a,b) \ge 0$; $d(a,b) = 0 \Leftrightarrow a = b$.
  (ii) $d(a,b) = d(b,a)$.
  (iii) $d(a,c) \le d(a,b) + d(b,c)$.

5. If X is any nonempty set, a function $d : X \times X \to \mathbb{R}$ with the properties (i)–(iii) of Exercise 4 is called a **metric** (or 'distance function') on X. These properties are called, respectively, *strict positivity, symmetry,* and the *triangle inequality.* Verify that the function $D : \mathbb{R} \times \mathbb{R} \to \mathbb{R}$ defined by the formula
$$D(a,b) = \frac{|a-b|}{1+|a-b|}$$
is also a metric on $\mathbb{R}$. {Hint: §1.2, Exercise 3.}

6. Let X be any nonempty set. For every pair of points x,y of X, define
$$d(x,y) = \begin{cases} 1 & \text{if } x \ne y, \\ 0 & \text{if } x = y. \end{cases}$$

Prove that d is a metric on X in the sense of Exercise 5 (it is called the *discrete metric* on X).

7. If d is a metric on the set X (Exercise 5), deduce from the triangle inequality that
$$|d(x,y) - d(x',y')| \le d(x,x') + d(y,y')$$
for all $x,y,x',y' \in X$.
  {Hint: $d(x,y) \le d(x,x') + d(x',y) \le d(x,x') + d(x',y') + d(y',y)$.}

CHAPTER 3

Sequences of Real Numbers, Convergence

The chapter begins with a discussion of sequences that culminates in the concept of convergence, the fundamental concept of analysis. The Weierstrass-Bolzano theorem (§3.5), nominally a theorem about bounded sequences, is in essence a property of closed intervals; Cauchy's criterion (§3.6) is a test for convergence, especially useful in the theory of infinite series (§10.1). The chapter concludes with a dissection of convergence into two more general limiting operations.

3.1. Bounded Sequences

For a review of sequences in general, see §2.5.

3.1.1. *Definition.* A sequence (x_n) of real numbers is said to be **bounded** if the set $\{ x_n : n \in \mathbb{P}\}$ is bounded in the sense of 1.3.1. A sequence that is not bounded is said to be **unbounded**.

3.1.2. *Remark.* A sequence (x_n) in $\mathbb{R}$ is bounded if and only if there exists a positive real number K such that $|x_n| \leq K$ for all n.

{Proof: If $a \leq x_n \leq b$ for all n and if $K = |a| + |b|$ (for example) then $|a| \leq K$ and $|b| \leq K$, thus

$$-K \leq -|a| \leq a \leq x_n \leq b \leq |b| \leq K$$

by (8) of 2.9.2, therefore $|x_n| \leq K$ by (9) of 2.9.2.}

3.1.3. *Examples.* Every constant sequence ($x_n = x$ for all n) is bounded. The sequence $x_n = (-1)^n$ is bounded.

33

3.1.4. *Example.* The sequence $x_n = n$ is unbounded. {For every real number K there exists, by the Archimedean property, a positive integer n such that $n = n1 > K$, thus the set of all x_n is not bounded above.}

3.1.5. **Theorem.** *If (x_n) and (y_n) are bounded sequences in $\mathbb{R}$, then the sequences $(x_n + y_n)$ and $(x_n y_n)$ are also bounded.*

Proof. If $|x_n| \leq K$ and $|y_n| \leq K'$ then $|x_n + y_n| \leq |x_n| + |y_n| \leq K + K'$ and $|x_n y_n| = |x_n| |y_n| \leq KK'$. $\Diamond$

Exercises

1. Show that if (x_n) and (y_n) are bounded sequences and $c \in \mathbb{R}$, then the sequences (cx_n) and $(x_n - y_n)$ are also bounded.

2. True or false (explain): If (x_n) and $(x_n y_n)$ are bounded, then (y_n) is also bounded.

3. Show that the sequence (x_n) defined by

$$x_n = 1 + \frac{1}{2} + \frac{1}{3} + \ldots + \frac{1}{n}$$

is unbounded. {Hint: $x_{2n} - x_n \geq 1/2$ for all n.}

4. Show that the sequence (x_n) defined recursively by $x_1 = 1$, $x_{n+1} = x_n + 1/x_n$ is unbounded.

{Hint: Assuming to the contrary that the sequence has an upper bound K, argue that $K - 1/K$ is also an upper bound.}

5. *Show that the sequence $x_n = \sin(n^2 + 1)$ is bounded.*[1]

6. Show that the sequence (x_n) defined by

$$x_n = 1 + \frac{1}{2!} + \frac{1}{3!} + \ldots + \frac{1}{n!}$$

is bounded above by 2. {Hint: $n! \geq 2^{n-1}$ for all n.}

7. Show that the sequence (a_n) defined by

$$a_n = \left(1 + \frac{1}{n}\right)^n$$

is bounded above by 3. {Hint: By the binomial formula,

$$a_n = \sum_{k=0}^{n} \binom{n}{k} \frac{1}{n^k};$$

show that the term for index k is $\leq 1/k!$, then cite Exercise 6.}

---

[1]Statements between split-level asterisks involve objects not yet constructed or properties not yet proved, but likely to be familiar from other courses.

3.2. Ultimately, Frequently

Let's pause to introduce some terminology useful for general sequences:

3.2.1. *Definition.* Let (x_n) be a sequence in a set X and let A be a subset of X.

(i) We say that $x_n \in A$ **ultimately** if x_n belongs to A from some index onward, that is, there is an index N such that $x_n \in A$ for all $n \geq N$; symbolically,

$$\exists N \ni n \geq N \implies x_n \in A.$$

{Equivalently, $\exists N \ni n > N \implies x_n \in A$, because $n > N$ means the same thing as $n \geq N + 1$.}

(ii) We say that $x_n \in A$ **frequently** if for every index N there is an index $n \geq N$ for which $x_n \in A$; symbolically,

$$(\forall N) \, \exists \, n \geq N \ni x_n \in A.$$

{Equivalently, $(\forall N) \, \exists \, n > N \ni x_n \in A.$}

3.2.2. *Example.* Let $x_n = 1/n$, let $\epsilon > 0$ and let $A = (0, \epsilon)$; then $x_n \in A$ ultimately.

{Proof: Choose an index N such that $1/N < \epsilon$ (2.1.2); then $n \geq N \implies 1/n \leq 1/N < \epsilon$.}

3.2.3. *Example.* For each positive integer n, let S_n be a statement (which may be either true or false). Let

$$A = \{n \in \mathbb{P} : S_n \text{ is true}\}.$$

We say that S_n is true *frequently* if $n \in A$ frequently, and that S_n is true *ultimately* if $n \in A$ ultimately. For example, $n^2 - 5n + 6 > 0$ ultimately (in fact, for $n \geq 4$); and n is frequently divisible by 5 (in fact, for $n = 5$, $n = 10$, $n = 15$, etc.).

There's an important logical relation between the two concepts in 3.2.1; to say that something *doesn't* happen ultimately means that its *negation* happens frequently:

3.2.4. Theorem. *With notations as in 3.2.1, one and only one of the following conditions holds:*
(1) $x_n \in A$ *ultimately;*
(2) $x_n \notin A$ *frequently.*

Proof. To say that (1) is false means that, whatever index N is proposed, the implication

$$n \geq N \implies x_n \in A$$

is false, so there must exist an index $n \geq N$ for which $x_n \notin A$; this is precisely the meaning of (2). $\Diamond$

For example (cf. 3.2.3) if (x_n) is a sequence in $\mathbb{R}$ then either $x_n < 5$ ultimately, or $x_n \geq 5$ frequently, but not both.

Exercises

1. For the sequence $x_n = n$, x_n is both frequently even and frequently odd; does this conflict with 3.2.4?

2. *True or false (explain):
(i) If $x_n = \sin(n\pi/2)$ then $x_n = 1$ frequently.
(ii) If $x_n = \sin(2\pi/n)$ then $x_n < 1$ ultimately.*

3. Let (A_n) be a sequence of subsets of a set X. Define

$$\limsup A_n = \{x \in X : x \in A_n \text{ frequently}\},$$
$$\liminf A_n = \{x \in X : x \in A_n \text{ ultimately}\}.$$

Prove:
(i) $\liminf A_n \subset \limsup A_n$.
(ii) $\liminf A_n = \bigcup_{n=1}^{\infty} \bigcap_{k=n}^{\infty} A_k$.
(iii) $(\limsup A_n)' = \liminf A_n'$ (' means complement).
(iv) $\limsup A_n = \bigcap_{n=1}^{\infty} \bigcup_{k=n}^{\infty} A_k$.

4. With notations as in Exercise 3, call (A_n) a *convergent* sequence of sets if $\liminf A_n$ and $\limsup A_n$ are equal; their common value A is then called the *limit* of the sequence, denoted $\lim A_n$, and the sequence is said to *converge* to A, written $A_n \to A$.

Prove: If (A_n) is an increasing sequence of sets with union A (concisely, $A_n \uparrow A$), then $A_n \to A$. State and prove the analogous proposition for decreasing sequences.

3.3. Null Sequences

3.3.1. *Definition.* A sequence (x_n) in $\mathbb{R}$ is said to be **null** if, for every positive real number ϵ, $|x_n| < \epsilon$ ultimately.

For example, the sequence $(1/n)$ is null (3.2.2).

The concept of convergence (§3.4) will be defined in such a way that the null sequences are precisely the sequences that converge to 0; thus the 'ultimately' concept is at the core of the concept of convergence.

Definition 3.3.1 can be expressed as follows: Given any $\epsilon > 0$ (no matter how small), the distance from x_n to the origin is ultimately smaller than ϵ

(in this sense x_n 'approaches' 0). A more informal[1] way to say it: x_n is arbitrarily small provided n is sufficiently large. {This is imprecise in the sense that "arbitrarily small" is understood to suggest that the degree of smallness is specified in advance, before any indices are selected; and "sufficiently large" is understood in the sense of "ultimately" (not merely "frequently").}

Various operations on null sequences lead to other null sequences (in particular, the example $x_n = 1/n$ generates an abundance of examples):

3.3.2. Theorem. *Let* (x_n) *and* (y_n) *be null sequences and let* $c \in \mathbb{R}$. *Then:*

(1) (x_n) *is bounded.*

(2) (cx_n) *is null.*

(3) $(x_n + y_n)$ *is null.*

(4) *If* (b_n) *is a bounded sequence then* $(b_n x_n)$ *is null.*

(5) *If* (z_n) *is a sequence such that* $|z_n| \leq |x_n|$ *ultimately, then* (z_n) *is also null.*

Proof. (1) Applying 3.3.1 with $\epsilon = 1$, we have $|x_n| < 1$ ultimately; let N be an index such that $|x_n| < 1$ for all $n > N$. If K is the larger of the numbers $1, |x_1|, |x_2|, \ldots, |x_N|$, then $|x_n| \leq K$ for every positive integer n, thus (x_n) is bounded.

(3) Let $\epsilon > 0$. Since (x_n) is null, there is an index N_1 such that

$$n \geq N_1 \implies |x_n| < \epsilon/2;$$

similarly, there is an index N_2 such that

$$n \geq N_2 \implies |y_n| < \epsilon/2,$$

so if N is the larger of N_1 and N_2 then

$$n \geq N \implies |x_n + y_n| \leq |x_n| + |y_n| < \epsilon/2 + \epsilon/2 = \epsilon.$$

This proves that $(x_n + y_n)$ is null.

(4) Let K be a positive real number such that $|b_n| \leq K$ for all n. Given any $\epsilon > 0$, choose an index N such that

$$n \geq N \implies |x_n| < \epsilon/K;$$

---

[1] In mathematics, 'formal' and 'rigorous' are code words for 'precision'. The aim of precision is to clarify; if an argument already appears to be clear then there is no *demand* for greater precision (though there may be a *need* for greater precision—indeed, greater precision of expression may reveal the argument, and even the asserted proposition, to be faulty). Since clarity is partly in the eyes of the beholder, standards of rigor can vary from generation to generation (and, at a given time, from one individual to another).

then

$$n \geq N \;\Rightarrow\; |b_n x_n| = |b_n|\,|x_n| \leq K|x_n| < \epsilon,$$

thus $(b_n x_n)$ is null.

(2) is a special case of (4).

(5) By assumption, there is an index N_1 such that

$$n \geq N_1 \;\Rightarrow\; |z_n| \leq |x_n|.$$

Given any $\epsilon > 0$, choose an index N_2 such that

$$n \geq N_2 \;\Rightarrow\; |x_n| < \epsilon;$$

if $N = \max\{N_1, N_2\}$ then

$$n \geq N \;\Rightarrow\; |z_n| \leq |x_n| < \epsilon,$$

thus (z_n) is null. $\Diamond$

As an example of (5), if $|x_n| \leq 1/n$ ultimately then (x_n) is null. Another important source of examples:

3.3.3. Theorem. *If* $a_n \uparrow a$ *or* $a_n \downarrow a$ *then the sequence* $(a_n - a)$ *is null.*

Proof. If $a_n \downarrow a$ then, by 2.5.6,

$$a_n - a = a_n + (-a) \downarrow a + (-a) = 0.$$

In particular, $a_n - a \geq 0$ and $\inf(a_n - a) = 0$. Given any $\epsilon > 0$, choose an index N such that $a_N - a < \epsilon$; then

$$n \geq N \;\Rightarrow\; |a_n - a| = a_n - a \leq a_N - a < \epsilon,$$

thus $(a_n - a)$ is null. The case that $a_n \uparrow a$ is deduced by applying the foregoing to $-a_n \downarrow -a$. $\Diamond$

3.3.4. Example. If $|x| < 1$ then the sequence (x^n) is null. {Proof: Writing $c = |x|$, we have $c^n \downarrow 0$ by 2.5.5; thus, the sequence $(|x^n|) = (c^n)$ is null (3.3.3), therefore so is the sequence (x^n).}

3.3.5. Example. Fix $x \in \mathbb{R}$ and let $x_n = x$ for all n. The constant sequence (x_n) is null if and only if $x = 0$. {Proof: The condition "$|x_n| < \epsilon$ ultimately" means $|x| < \epsilon$; if this happens for every $\epsilon > 0$, then $x = 0$ (for, if $x \neq 0$ then $\epsilon = \frac{1}{2}|x|$ is an embarrassment).}

Exercises

1. Prove: If (x_n) and (y_n) are null then so is $(x_n y_n)$.

2. Prove that (x_n) is null if and only if (x_n^2) is null. (Generalization?)

3. True or false (explain): If (x_n) and (y_n) are sequences in $\mathbb{R}$ such that $(x_n y_n)$ is null, then either (x_n) or (y_n) is null.

4. Let $x_n = \sqrt{n+1} - \sqrt{n}$. True or false (explain): (x_n) is null. {Hint: Consider $y_n = \sqrt{n+1} + \sqrt{n}$ and calculate $x_n y_n$.}

5. Show that a sequence (x_n) in $\mathbb{R}$ is *not* null if and only if there exists a positive number ϵ such that $|x_n| \geq \epsilon$ frequently.

6. *True or false (explain): The sequence $x_n = n[1 - \cos(1/n)]$ is null.*

7. If $|a| < 1$ prove that the sequence $a_n = n|a|^n$ is null. {Hint: Look at a_{n+1}/a_n.}

8. Show that the sequence $a_n = \sqrt{n^2 + 1} - n$ is null.

9. Let (a_n) be a null sequence in $\mathbb{R}$, let $\sigma : \mathbb{P} \to \mathbb{P}$ be injective and define $b_n = a_{\sigma(n)}$ (in other words, the n'th term of the new sequence is the $\sigma(n)$'th term of the old sequence). Prove that (b_n) is also a null sequence; in particular, every rearrangement of a null sequence is null.
{Hint: If N is a positive integer and M is an integer larger than every element of $\sigma^{-1}(\{1, 2, \ldots, N\})$, then $m \geq M \Rightarrow \sigma(m) \geq N$.}

10. If (x_n) is a null sequence and $y_n = (x_1 + \ldots + x_n)/n$ (the average of the first n terms), then (y_n) is also a null sequence.
{Hint: Suppose $|x_n| \leq \epsilon$ for all $n > N$. If $n > N$, then

$$y_n = y_N(N/n) + (x_{N+1} + \ldots + x_n)/n \, ;$$

estimate the second term on the right, noting that the number of terms in the numerator is $< n$.}

11. If $p : \mathbb{R} \to \mathbb{R}$ is a polynomial function without constant term and (x_n) is a null sequence, then $(p(x_n))$ is null.

3.4. Convergent Sequences

Taking a cue from Theorem 3.3.3, we make the following definition:

3.4.1. *Definition.* A sequence (a_n) in $\mathbb{R}$ is said to be **convergent** in $\mathbb{R}$ if there exists a real number a such that the sequence $(a_n - a)$ is null, and **divergent** if no such number exists.

Such a number a (if it exists) is unique. {Proof: Suppose that both $(a_n - a)$ and $(a_n - b)$ are null. Let $x_n = (a_n - b) - (a_n - a) = a - b$. Being the difference of null sequences, (x_n) is null (3.3.2), therefore $a - b = 0$ (3.3.5).}

3.4.2. *Definition.* With notations as in 3.4.1, the number a is called the **limit** of the convergent sequence (a_n), and the sequence is said to **converge** to a; this is expressed by writing

$$\lim_{n\to\infty} a_n = a,$$

or

$$a_n \to a \quad \text{as} \quad n \to \infty,$$

more concisely, $\lim a_n = a$ or $a_n \to a$.

3.4.3. *Example.* If $a_n \uparrow a$ or $a_n \downarrow a$ then $a_n \to a$ (3.3.3), thus every bounded monotone sequence is convergent (2.5.3).

3.4.4. *Example.* $a_n \to 0 \Leftrightarrow (a_n)$ is null. {Since $a_n - 0 = a_n$, this is immediate from the definitions.}

3.4.5. *Example.* $a_n \to a \Leftrightarrow a_n - a \to 0$. {Immediate from 3.4.4 and the definitions.}

3.4.6. *Example.* $x^n \to 0 \Leftrightarrow |x| < 1$. {Proof: If $|x| < 1$ then $x^n \to 0$ by 3.3.4; if $|x| \geq 1$ then $|x^n| = |x|^n \geq 1$ for all n, therefore (x^n) is not null.}

3.4.7. *Example.* If $|x| < 1$ and $a_n = 1 + x + x^2 + \ldots + x^{n-1}$, then $a_n \to 1/(1-x)$; for,

$$a_n - \frac{1}{1-x} = \frac{1-x^n}{1-x} - \frac{1}{1-x} = \frac{-1}{1-x} \cdot x^n$$

is a constant multiple of a null sequence (3.4.6), hence is null (3.3.2).

The basic properties of convergence are as follows:

3.4.8. **Theorem.** *Let (a_n), (b_n) be convergent sequences in $\mathbb{R}$, say $a_n \to a$ and $b_n \to b$, and let $c \in \mathbb{R}$. Then:*
(1) *(a_n) is bounded,*
(2) *$ca_n \to ca$,*
(3) *$a_n + b_n \to a + b$,*
(4) *$a_n b_n \to ab$,*
(5) *$|a_n| \to |a|$.*
(6) *If $b \neq 0$ then $|b_n|$ is ultimately bounded away from 0, in the sense that there exists an $r > 0$ (for example, $r = \frac{1}{2}|b|$) such that $|b_n| \geq r$ ultimately.*
(7) *If b and the b_n are all nonzero, then $a_n/b_n \to a/b$.*
(8) *If $a_n \leq b_n$ for all n, then $a \leq b$.*

Proof. (1) $(a_n - a)$ is null, therefore bounded (3.3.2), and $a_n = (a_n - a) + a$ shows that (a_n) is the sum of two bounded sequences.

(2) $ca_n - ca = c(a_n - a)$ is a scalar multiple of a null sequence, so it is null by 3.3.2.

(3) $(a_n + b_n) - (a + b) = (a_n - a) + (b_n - b)$ is the sum of two null sequences, therefore is null (3.3.2).

(4) $a_n b_n - ab = a_n(b_n - b) + (a_n - a)b$; since (a_n) is bounded and $(b_n - b)$, $(a_n - a)$ are null, it follows from 3.3.2 that $(a_n b_n - ab)$ is null.

(5) By 2.9.2, $||a_n| - |a|| \leq |a_n - a|$, where $(a_n - a)$ is null, so $(|a_n| - |a|)$ is null by 3.3.2.

(6) Let $r = \frac{1}{2}|b|$ and choose an integer N such that

$$n \geq N \implies |b_n - b| \leq r.$$

Then, for all $n \geq N$,

$$2r = |b| = |(b - b_n) + b_n| \leq |b - b_n| + |b_n| \leq r + |b_n|,$$

so $|b_n| \geq r$.

(7) From (6), it follows that the sequence $(1/b_n)$ is bounded; thus

$$\frac{1}{b_n} - \frac{1}{b} = \frac{1}{b_n b}(b - b_n)$$

is the product of a bounded sequence and a null sequence, therefore is null. Thus $1/b_n \to 1/b$, therefore

$$a_n/b_n = a_n(1/b_n) \to a(1/b) = a/b$$

by (4).

(8) Let $c_n = b_n - a_n$ and $c = b - a$; then $c_n \geq 0$ and $c_n \to c$, and our problem is to show that $c \geq 0$. By (5), $|c_n| \to |c|$, that is, $c_n \to |c|$; already $c_n \to c$, so $c = |c| \geq 0$ by the uniqueness of limits. ◊

3.4.9. **Theorem.** *If* $A \subset \mathbb{R}$ *is nonempty and bounded above, and if* $M = \sup A$, *then there exists a sequence* (x_n) *in* A *such that* $x_n \to M$.

Proof. For each positive integer n, choose $x_n \in A$ so that $M - 1/n < x_n \leq M$ (1.3.5). ◊

Similarly, if $A \subset \mathbb{R}$ is nonempty and bounded below, then $\inf A$ is the limit of a sequence in A.

Exercises

1. If $a_n \to a$, $b_n \to b$ and $a_n \leq b_n$ frequently, show that $a \leq b$.

2. Suppose $a_n \to a$. Define

$$s_n = \frac{1}{n} \sum_{k=1}^{n} a_k.$$

Prove that $s_n \to a$. {Hint: §3.3, Exercise 10.}

3. Show that the sequence

$$a_n = \left(1 - \frac{1}{2}\right)\left(1 - \frac{1}{3}\right)\cdots\left(1 - \frac{1}{n+1}\right)$$

is convergent. {Hint: Calculate a_n for the first few values of n.}

4. Let a and b be fixed real numbers and define a sequence (x_n) as follows: $x_1 = a$, $x_2 = b$ and, for $n > 2$, x_n is the average of the preceding two terms, that is, $x_n = \frac{1}{2}(x_{n-2} + x_{n-1})$.

(i) Draw a picture to see what is going on. {Assume $a < b$ and think midpoints.}

(ii) Prove (by induction) that

$$x_{n+1} - x_n = \left(-\frac{1}{2}\right)^{n-1}(b - a).$$

(iii) Show that

$$x_{n+1} - x_1 = \left[\sum_{k=1}^{n}\left(-\frac{1}{2}\right)^{k-1}\right](b - a).$$

(iv) Prove that $x_n \to (a + 2b)/3$.

5. Find $\lim a_n$ for the following sequences:

(i) $a_n = \frac{1}{n^2} + \frac{2}{n^2} + \ldots + \frac{n}{n^2}$.

(ii) $a_n = \frac{(1/n)^2 + (2/n)^2 + \ldots + (n/n)^2}{n}$.

6. Given the polynomial functions

$$p(x) = 2x^4 + x^2 + 3x + 1, \quad q(x) = 3x^4 + x^3 + 2x + 3,$$

define

$$a_n = \frac{p(n)}{q(n)} \quad (n = 1, 2, 3, \ldots).$$

Prove that $a_n \to 2/3$.

7. Prove that the sequence

$$a_n = \frac{1}{n+1} + \frac{1}{n+2} + \ldots + \frac{1}{n+n}$$

is convergent, to a limit ≤ 1. {Hint: Show that the sequence is increasing and bounded above by 1.}

8. Prove that the sequence (a_n) defined recursively by $a_1 = 1$ and $a_{n+1} = a_n + 1/a_n$ ($n = 1, 2, 3, \ldots$) is not bounded. {Hint: 3.4.3.}

9. Prove that the sequence

$$a_n = \left(1 + \frac{1}{n}\right)^n$$

is convergent. {Hint: §1.2, Exercise 20 and §3.1, Exercise 7.}

10. Let $a_n = n!/n^n$. Show that $a_n \to 0$. {Hint: Show that $a_n \downarrow$ and that a_{n+1}/a_n has a limit < 1.}

11. Let (a_n) be a convergent sequence in $\mathbb{R}$, let $\sigma : \mathbb{P} \to \mathbb{P}$ be injective and define $b_n = a_{\sigma(n)}$. Prove that (b_n) is also convergent and that $\lim b_n = \lim a_n$. {Hint: §3.3, Exercise 9.}

12. Let X be a nonempty set and let d be a metric on X (§2.9, Exercise 5).
  (i) If (x_n) is a sequence in X and if x, y are elements of X such that $d(x_n, x) \to 0$ and $d(x_n, y) \to 0$, prove that $x = y$. {Hint: $d(x,y) \leq d(x, x_n) + d(x_n, y)$.}
  (ii) A sequence (x_n) in X is said to be *convergent* if there exists a point $x \in X$ such that $d(x_n, x) \to 0$. By (i), such a point x is unique; it is called the *limit* of the sequence (x_n), written $x = \lim x_n$, and the sequence is said to *converge* to x, written $x_n \to x$.
  Prove: If $x_n \to x$ and $y_n \to y$ then $d(x_n, y_n) \to d(x, y)$. {Hint: §2.9, Exercise 7.}
  (iii) If d is the discrete metric on X (§2.9, Exercise 6), a sequence in X is convergent if and only if it is ultimately constant. {Hint: If $d(x_n, x) \to 0$ then $d(x_n, x) < 1$ ultimately; the only available distances are 0 and 1.}

13. Let d be a metric on the set X, and let D be the metric on X defined by the formula $D = d/(1 + d)$ (cf. §2.9, Exercise 5).
  (i) Show that $d = D/(1 - D)$.
  (ii) Prove that $d(x_n, x) \to 0 \Leftrightarrow D(x_n, x) \to 0$; in other words, convergence for the metric d means the same thing as convergence for the metric D.

14. Let $[a_n, b_n]$ be a nested sequence of closed intervals such that $b_n - a_n \downarrow 0$ and let (x_n) be a sequence such that $x_n \in [a_n, b_n]$ for all n. Prove that (x_n) is convergent.

3.5. Subsequences, Weierstrass-Bolzano Theorem

Given a sequence (x_n), there are various ways of forming 'subsequences': for example, take every other term,

$$x_1, x_3, x_5, \ldots ;$$

or, take all of the terms from some index onward,

$$x_6, x_7, x_8, \dots ;$$

or take all terms for which the index is a prime number,

$$x_2, x_3, x_5, x_7, x_{11}, \dots .$$

The general idea is that one is free to discard any terms, as long as infinitely many terms remain. The formal definition is as follows:

3.5.1. *Definition.* Let (x_n) be any sequence. Choose a strictly increasing sequence of positive integers

$$n_1 < n_2 < n_3 < \dots$$

and define $y_k = x_{n_k}$ ($k = 1, 2, 3, \dots$). One calls (y_k) a **subsequence** of (x_n) . This is also expressed by saying that $x_{n_1}, x_{n_2}, x_{n_3}, \dots$ is a subsequence of $x_1, x_2, x_3, \dots$, or that (x_{n_k}) is a subsequence of (x_n) .

Forming a subsequence amounts to choosing a sequence of indices; what is essential is that the chosen indices must march steadily to the right—no stuttering, no dropping back.

3.5.2. *Remark.* Let (n_k) be a strictly increasing sequence of positive integers:

$$n_1 < n_2 < n_3 < \dots .$$

For every positive integer N, there exists a positive integer k such that $n_k > N$ (therefore $n_j > N$ for all $j \geq k$). {Proof: It clearly suffices to show that $n_k \geq k$ for all positive integers k. This is obvious for $k = 1$. Assuming inductively that $n_k \geq k$, we have $n_{k+1} \geq n_k + 1 \geq k + 1$.}

3.5.3. *Remark.* Here's a useful perspective on subsequences. A sequence (x_n) in a set X can be thought of as a function $f : \mathbb{P} \to X$, where $f(n) = x_n$. A subsequence of (x_n) is obtained by specifying a strictly increasing function $\sigma : \mathbb{P} \to \mathbb{P}$ and taking the composite function $f \circ \sigma$ (Figure 3):

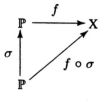

FIGURE 3

Thus, writing $n_k = \sigma(k)$, we have $(f \circ \sigma)(k) = f(\sigma(k)) = f(n_k) = x_{n_k}$.

3.5.4. *Remark.* An application of the preceding remark: If (y_k) is a subsequence of (x_n), then every subsequence of (y_k) is also a subsequence of (x_n). The crux of the matter is that if $\sigma : \mathbb{P} \to \mathbb{P}$ and $\tau : \mathbb{P} \to \mathbb{P}$ are strictly increasing, then so is $\sigma \circ \tau : \mathbb{P} \to \mathbb{P}$.

{In detail, suppose that $f : \mathbb{P} \to X$ defines the sequence (x_n), that is, $f(n) = x_n$, and that $\sigma : \mathbb{P} \to \mathbb{P}$ defines the subsequence (y_k), that is, $y_k = f(\sigma(k))$. Write $g = f \circ \sigma$; then $g : \mathbb{P} \to X$ with $g(k) = y_k$. Suppose (z_i) is a subsequence of (y_k), say defined by $\tau : \mathbb{P} \to \mathbb{P}$, so that $z_i = g(\tau(i))$. Then $g \circ \tau = (f \circ \sigma) \circ \tau = f \circ (\sigma \circ \tau)$, so (z_i) is defined by the strictly increasing function $\sigma \circ \tau : \mathbb{P} \to \mathbb{P}$.}

The following properties of subsequences will be used on many occasions:

3.5.5. **Theorem.** *Let* (a_n) *be a sequence in* $\mathbb{R}$ *and let* (a_{n_k}) *be a subsequence of* (a_n).

(1) *If* (a_n) *is bounded, then so is* (a_{n_k}).

(2) *If* (a_n) *is null, then so is* (a_{n_k}).

(3) *If* (a_n) *is convergent, then so is* (a_{n_k}); *more precisely, if* $a_n \to a$ *as* $n \to \infty$ *then also* $a_{n_k} \to a$ *as* $k \to \infty$.

(4) *If* $a_n \uparrow a$ *then also* $a_{n_k} \uparrow a$, *and similarly for decreasing sequences.*

Proof. (1) If $|a_n| \leq K$ for all n, then in particular $|a_{n_k}| \leq K$ for all k.

(2) Write $b_k = a_{n_k}$ ($k = 1, 2, 3, \ldots$). Let $\epsilon > 0$. By assumption, there is an index N such that $|a_n| < \epsilon$ for all $n \geq N$. Choose k so that $n_k \geq N$ (3.5.2); then

$$j \geq k \;\Rightarrow\; n_j \geq n_k \geq N \;\Rightarrow\; |a_{n_j}| < \epsilon.$$

Thus (b_k) is null.

(3) By assumption, $(a_n - a)$ is null, therefore so is its subsequence $(a_{n_k} - a)$, consequently $a_{n_k} \to a$.

(4) If $a_n \uparrow a$ the subsequence (a_{n_k}) is certainly increasing and bounded above (by a); writing

$$b = \sup\{a_{n_k} : k \in \mathbb{P}\},$$

we know that $b \leq a$ and we have to show that $b = a$. Given any positive integer n, there is a k such that $n_k > n$ (3.5.2), therefore

$$a_n \leq a_{n_k} \leq b;$$

thus $a_n \leq b$ for every positive integer n, therefore $a \leq b$.

If $a_n \downarrow a$ then $-a_n \uparrow -a$, therefore $-a_{n_k} \uparrow -a$ by the foregoing, consequently $a_{n_k} \downarrow a$. ◊

Subsequences often arise in situations like the following:

3.5.6. Theorem. *Let (x_n) be a sequence and let (P) be a property that a term x_n may or may not have. Then the following conditions are equivalent:*

(a) x_n *has property (P) frequently;*

(b) *there exists a subsequence (x_{n_k}) of (x_n) such that every x_{n_k} has property (P).*

Proof. Let $A = \{n \in \mathbb{P} : x_n \text{ has property (P)}\}$.

(a) $\Rightarrow$ (b): By assumption, $n \in A$ frequently (3.2.3). Choose $n_1 \in A$. Choose $n_2 \in A$ so that $n_2 > n_1$ (3.2.1). Choose $n_3 \in A$ so that $n_3 > n_2$, and so on. The subsequence $x_{n_1}, x_{n_2}, x_{n_3}, \ldots$ constructed in this way has the desired property.

(b) $\Rightarrow$ (a): By assumption, $n_k \in A$ for all k. Given any index N, the claim is that A contains an integer $n \geq N$. Indeed, $n_k > N$ for some k (3.5.2). $\Diamond$

Here is the kind of application for which the preceding theorem is useful:

3.5.7. Example. Suppose we're trying to show that $|a_n - a| < \epsilon$ ultimately. The alternative (3.2.4) is that $|a_n - a| \geq \epsilon$ frequently, in other words (3.5.6) $|a_{n_k} - a| \geq \epsilon$ for some subsequence (a_{n_k}) of (a_n); it may be more convenient to prove that the alternative is false.

The next theorem is, at first glance, surprising:

3.5.8. Theorem. *Every sequence in $\mathbb{R}$ has a monotone subsequence.*

Proof. Assuming (a_n) is any sequence of real numbers, we seek a subsequence (a_{n_k}) that is either increasing or decreasing (2.5.2).

Call a positive integer n a *peak point* for the sequence if $a_n \geq a_k$ for all $k \geq n$. {Think of the sequence as a function $f : \mathbb{P} \to \mathbb{R}$, $f(n) = a_n$. For n to be a peak point means that no point of the graph of f from n onward is higher than (n, a_n); so to speak, 'a_n is not topped to the right'.} There are two possibilities:

(1) Suppose first that n is frequently a peak point. If $n_1 < n_2 < n_3 < \ldots$ are peak points, then the subsequence (a_{n_k}) is decreasing. For, $a_{n_1} \geq a_{n_2}$ (because n_1 is a peak point), $a_{n_2} \geq a_{n_3}$ (because n_2 is a peak point), etc.

(2) The alternative is that, from some index N onward, n is not a peak point. Let $n_1 = N$. Since n_1 is not a peak point, a_{n_1} is 'topped to the right': there is an index $n_2 > n_1$ such that $a_{n_2} > a_{n_1}$. But n_2 isn't a peak point either, so there is an $n_3 > n_2$ such that $a_{n_3} > a_{n_2}$. Continuing in this way, we obtain an increasing subsequence of (a_n). $\Diamond$

The theorem has an equally surprising (and famous) consequence:

3.5.9. Theorem. (Weierstrass–Bolzano theorem) *Every bounded sequence in $\mathbb{R}$ has a convergent subsequence.*

Proof.[1] Let (a_n) be a bounded sequence of real numbers. By the preceding theorem, (a_n) has a monotone subsequence (a_{n_k}). Suppose, for example, that (a_{n_k}) is increasing; it is also bounded, so $a_{n_k} \uparrow a$ for a suitable real number a (2.5.3), and $a_{n_k} \to a$ by 3.4.3. ◇

The Weierstrass-Bolzano theorem can be reformulated as a theorem about closed intervals:

3.5.10. Corollary. *In a closed interval $[a, b]$, every sequence has a subsequence that converges to a point of the interval.*

Proof. Suppose $x_n \in [a, b]$ $(n = 1, 2, 3, \ldots)$. By the theorem, some subsequence is convergent to a point of $\mathbb{R}$, say $x_{n_k} \to x$; since $a \leq x_{n_k} \leq b$ for all k, it follows that $a \leq x \leq b$ (3.4.8), thus $x \in [a, b]$. ◇

Exercises

1. Let (a_n) be a sequence in $\mathbb{R}$. Prove that (a_n) is unbounded if and only if there exists a subsequence (a_{n_k}) such that $|a_{n_k}| \geq k$ for all k.

2. Let (a_n) be a sequence in $\mathbb{R}$. Prove that the following conditions are equivalent:
  (a) (a_n) is divergent;
  (b) for every $a \in \mathbb{R}$ there exist an $\epsilon > 0$ and a subsequence (a_{n_k}) such that $|a_{n_k} - a| \geq \epsilon$ for all k.

3. *True or false (explain): The sequence $a_n = \sin n$ has a convergent subsequence.*

4. True or false (explain): If (a_n) is any sequence in $\mathbb{R}$, then the sequence
$$x_n = \frac{a_n}{1 + |a_n|}$$
has a convergent subsequence.

5. Construct a sequence that has a subsequence converging to 0 and another converging to 1. (Generalization?)

6. Let (a_n) be a sequence in $\mathbb{R}$ that is bounded but not convergent. Show that there exist two subsequences converging to different limits. {Hint: Let a be a real number such that some subsequence of (a_n) converges to a, then apply Exercise 2.}

7. Prove that the following conditions on a sequence (a_n) in $\mathbb{R}$ are equivalent:
  (a) (a_n) is bounded;

---

[1]This elegant argument is taken from the book of W. Maak [*An introduction to modern calculus*, Holt, 1963], p. 30.

(b) every subsequence of (a_n) has a convergent subsequence.
{Hint: Exercise 1.}

8. If X is a nonempty set, and d is a metric on X (§2.9, Exercise 5), the pair (X, d) is called a *metric space*. A metric space is said to be *compact* if every sequence in the space has a convergent subsequence (cf. §3.4, Exercise 12).

(i) Let X be an interval in $\mathbb{R}$ with endpoints a and b (1.3.2), where $a < b$, and let $d(x, y) = |x - y|$ be the usual metric on X. Prove: X is compact if and only if $X = [a, b]$. {Hint: The "if" part is covered by 3.5.10. Show that the other three kinds of intervals with endpoints a and b are not compact.}

(ii) If (X, d) is a compact metric space, show that the subset $\{d(x, y) : x, y \in X\}$ of $\mathbb{R}$ is bounded. {Hint: For every pair of sequences (x_n) and (y_n) in X, the sequence of real numbers $d(x_n, y_n)$ has a convergent subsequence.}

9. Prove that if (a_n) is a monotone sequence in $\mathbb{R}$ that has a bounded subsequence, then (a_n) is convergent.

10. If (a_n) is a sequence in $\mathbb{R}$ and $a \in \mathbb{R}$, the following conditions are equivalent:
(a) $(\forall \epsilon > 0)$ $|a_n - a| < \epsilon$ frequently;
(b) there exists a subsequence of (a_n) converging to a.
If the word "frequently" in (a) is replaced by "ultimately", how must (b) be changed so as to remain equivalent to (a)?

3.6. Cauchy's Criterion for Convergence

The criterion for a monotone sequence to converge is that it be bounded (3.4.3, 3.4.8). Cauchy's criterion for convergence applies to sequences that are not necessarily monotone:

3.6.1. **Theorem.** (Cauchy's criterion) *For a sequence (a_n) in $\mathbb{R}$, the following conditions are equivalent:*
(a) *(a_n) is convergent;*
(b) *for every $\epsilon > 0$, there is an index N such that $|a_m - a_n| < \epsilon$ whenever m and n are $\geq N$; in symbols,*

$$(\forall \, \epsilon > 0) \, \exists N \ni m, n \geq N \Rightarrow |a_m - a_n| < \epsilon.$$

Proof. (a) $\Rightarrow$ (b): Say $a_n \to a$. If $\epsilon > 0$ then $|a_n - a| < \epsilon/2$ ultimately, say for $n \geq N$; if both m and n are $\geq N$ then, by the triangle inequality,

$$|a_m - a_n| = |(a_m - a) + (a - a_n)|$$
$$\leq |a_m - a| + |a - a_n| < \epsilon/2 + \epsilon/2 = \epsilon.$$

(b) $\Rightarrow$ (a): Assuming (b), let's show first that the sequence (a_n) is bounded. Choose an index M such that $|a_m - a_n| < 1$ for all $m, n \geq M$. Then, for all $n \geq M$,

$$|a_n| = |(a_n - a_M) + a_M| \leq |a_n - a_M| + |a_M| < 1 + |a_M|,$$

therefore the sequence (a_n) is bounded; explicitly, if

$$r = \max\{|a_1|, |a_2|, \ldots, |a_{M-1}|, 1 + |a_M|\},$$

then $|a_n| \leq r$ for all n.

By the Weierstrass-Bolzano theorem, (a_n) has a convergent subsequence, say $a_{n_k} \to a$ (3.5.9); we will show that $a_n \to a$.

Let $\epsilon > 0$. By hypothesis, there is an index N such that

(*) $m, n \geq N \Rightarrow |a_m - a_n| < \epsilon/2$.

Since $a_{n_k} \to a$, there is an index K such that

(**) $k \geq K \Rightarrow |a_{n_k} - a| < \epsilon/2$.

Choose an index k such that $n_k \geq N$ (3.5.2); while we're at it, we can require that $k \geq K$. Then, for all $n \geq N$,

$$|a_n - a| \leq |a_n - a_{n_k}| + |a_{n_k} - a| < \epsilon/2 + \epsilon/2$$

by (*) and (**). This shows that the sequence $(a_n - a)$ is null, so $a_n \to a$ (3.4.1). $\diamond$

The interest of Cauchy's criterion is that the condition in (b) can often be verified without any knowledge as to the value of the *limit* of the sequence (a virtue shared by the convergence criterion for monotone sequences mentioned at the beginning of the section).

Exercises

1. Let (a_n) be a sequence in $\mathbb{R}$ and define

$$s_n = a_1 + \ldots + a_n,$$
$$t_n = |a_1| + \ldots + |a_n|.$$

Prove: If (t_n) is convergent, then so is (s_n). {Hint: $|s_m - s_n| \leq |t_m - t_n|$ by the triangle inequality.}

2. Let (A_n) be a sequence of nonempty subsets of $\mathbb{R}$ such that
(i) $A_1 \supset A_2 \supset A_3 \supset \ldots$, and

(ii) $|x - y| \leq 1/n$ for all $x, y \in A_n$.

Let (a_n) be a sequence in $\mathbb{R}$ such that $a_n \in A_n$ for all n. Prove that (a_n) is convergent.

3. Let (X, d) be a metric space (§3.5, Exercise 8). A sequence (x_n) in X is said to be *Cauchy* (for the metric d) if, for every $\epsilon > 0$, there is an index N such that $m, n \geq N \Rightarrow d(x_m, x_n) < \epsilon$.

(i) If a sequence (x_n) in X is convergent (§3.4, Exercise 12) then it is Cauchy.

(ii) If X is the interval $(0, 1]$ in $\mathbb{R}$, with the usual metric $d(x, y) = |x - y|$, then the sequence $(1/n)$ is Cauchy but not convergent in X.

(iii) If d is the discrete metric on X then every Cauchy sequence in X is convergent. {Hint: Cf. §3.4, Exercise 12, (iii).}

4. A metric space (X, d) is said to be *complete* if every Cauchy sequence in X is convergent in X (cf. Exercise 3). Prove: A closed interval $[a, b]$ in $\mathbb{R}$, with the metric $d(x, y) = |x - y|$, is a complete metric space. {Hint: Cf. the proof of 3.5.10.}

5. Let (X, d) be a metric space and let D be the metric on X defined by the formula $D = d/(1 + d)$ (cf. §3.4, Exercise 13). Prove that (X, d) is complete (in the sense of Exercise 4) if and only if (X, D) is complete.

6. Prove that every compact metric space (§3.5, Exercise 8) is complete in the sense of Exercise 4.

3.7. Limsup and Liminf of a Bounded Sequence

For a monotone sequence, what does it take to be convergent? The answer is that it has to be bounded (3.4.3).

Let's turn the tables: For a bounded sequence, what does it take to be convergent? The answer given in this section is based on an analysis of two monotone sequences derived from the sequence; in effect, the question of general convergence is reformulated in terms of monotone convergence.

3.7.1. *Notations.* Let (a_n) be a bounded sequence in $\mathbb{R}$, say $|a_n| \leq K$ for all n. For each n, let A_n be the set of all terms from n onward,

$$A_n = \{a_n, a_{n+1}, a_{n+2}, \ldots\} = \{a_k : k \geq n\};$$

then A_n is bounded, indeed $A_n \subset [-K, K]$, and we may define

$$b_n = \sup A_n = \sup_{k \geq n} a_k, \quad c_n = \inf A_n = \inf_{k \geq n} a_k.$$

This produces two sequences (b_n) and (c_n), and it is clear that $c_n \leq b_n$ for all n. These sequences are also bounded; indeed,

$$-K \leq c_n \leq b_n \leq K$$

for all n. Moreover, (c_n) is increasing and (b_n) is decreasing; for, $A_n \supset A_{n+1}$, therefore

$$c_n = \inf A_n \le \inf A_{n+1} = c_{n+1},$$
$$b_n = \sup A_n \ge \sup A_{n+1} = b_{n+1};$$

thus, writing

$$c = \sup c_n = \sup\{c_n : n \in \mathbb{P}\},$$
$$b = \inf b_n = \inf\{b_n : n \in \mathbb{P}\},$$

we have $c_n \uparrow c$ and $b_n \downarrow b$.

3.7.2. *Definition.* With the above notations, b is called the **limit superior** of the bounded sequence (a_n), written

$$\limsup a_n = b = \inf_{n \ge 1} b_n = \inf_{n \ge 1}\left(\sup_{k \ge n} a_k\right),$$

and c is called the **limit inferior** of the sequence (a_n), written

$$\liminf a_n = c = \sup_{n \ge 1} c_n = \sup_{n \ge 1}\left(\inf_{k \ge n} a_k\right).$$

3.7.3. *Examples.* (i) For the sequence

$$1, -1, 1, -1, \ldots,$$

$A_n = \{-1, 1\}$ for all n, so $b_n = 1$ and $c_n = -1$ for all n, therefore $b = 1$ and $c = -1$.

(ii) For the sequence

$$1, -1, 1, 1, 1, \ldots,$$

$A_n = \{1\}$ for $n \ge 3$, so $b_n = c_n = 1$ for $n \ge 3$, therefore $b = c = 1$.

(iii) For the sequence

$$\frac{1}{2}, \frac{2}{3}, \frac{1}{3}, \frac{3}{4}, \frac{1}{4}, \frac{4}{5}, \ldots, \frac{1}{k}, \frac{k}{k+1}, \ldots,$$

$b_n = 1$ and $c_n = 0$ for all n, therefore $b = 1$ and $c = 0$.

3.7.4. **Theorem.** *For every bounded sequence (a_n) in $\mathbb{R}$, $\liminf a_n \le \limsup a_n$.*

Proof. In the preceding notations, the problem is to show that $c \le b$. By 3.4.3, we have $c_n \to c$ and $b_n \to b$; since $c_n \le b_n$ for all n, it follows that $c \le b$ (3.4.8). ∎

{Alternative proof: If m and n are any two positive integers and $p = \max\{m, n\}$, then $m \leq p$ and $n \leq p$, therefore

$$c_m \leq c_p \leq b_p \leq b_n .$$

This shows that each c_m is a lower bound for all the b_n, so $c_m \leq$ GLB $b_n = b$; then b is an upper bound for all the c_m, so $c =$ LUB $c_m \leq b$.} ◊

What does this have to do with convergence? Everything:

3.7.5. Theorem. *For a sequence* (a_n) *in* $\mathbb{R}$, *the following conditions are equivalent*:
(a) (a_n) *is convergent*;
(b) (a_n) *is bounded and* $\liminf a_n = \limsup a_n$.
For such a sequence, $\lim a_n = \liminf a_n = \limsup a_n$.

Proof. (a) ⇒ (b): If $a_n \to a$ then (a_n) is bounded (3.4.8) and our problem is to show that, in the notations of 3.7.2, $c = b = a$.

Let $\epsilon > 0$. Choose an index N such that $|a_n - a| \leq \epsilon$ for all $n \geq N$; then, for all $n \geq N$,

$$-\epsilon \leq a_n - a \leq \epsilon ,$$

that is, $a - \epsilon \leq a_n \leq a + \epsilon$. This shows (in the notations of 3.7.1) that

$$A_N \subset [a - \epsilon, a + \epsilon] ,$$

consequently

$$a - \epsilon \leq c_N \leq b_N \leq a + \epsilon ;$$

but $c_N \leq c \leq b \leq b_N$, so

$$a - \epsilon \leq c \leq b \leq a + \epsilon .$$

In particular,

$$a - 1/n \leq c \leq b \leq a + 1/n$$

for every positive integer n; since $1/n \to 0$, it follows from 3.4.8 that $a \leq c \leq b \leq a$, thus $a = c = b$.

(b) ⇒ (a): Assuming (a_n) is bounded, define b and c as in 3.7.1. {Later on, assuming $b = c$, we will show that (a_n) is convergent.}

Let $\epsilon > 0$. Since $b =$ GLB b_n is the *greatest* lower bound of the b_n and since $b + \epsilon > b$, $b + \epsilon$ can't be a lower bound for the b_n; thus $b + \epsilon$ is not $\leq$ every b_n, so $b + \epsilon > b_N$ for some N, that is,

$$b + \epsilon > \sup\{a_n : n \geq N\} .$$

Then $n \geq N \Rightarrow a_n < b + \epsilon$. We have shown that

$$(\forall \, \epsilon > 0) \ a_n < b + \epsilon \ \text{ultimately} .$$

A similar argument shows that

$$(\forall\, \epsilon > 0) \quad c - \epsilon < a_n \quad \text{ultimately} .$$

Combining these two results, we have

$$(\forall\, \epsilon > 0) \quad c - \epsilon < a_n < b + \epsilon \quad \text{ultimately} .$$

It follows that if $c = b$ and a denotes the common value of c and b, then

$$(\forall\, \epsilon > 0) \quad |a_n - a| < \epsilon \quad \text{ultimately} ;$$

in other words, $a_n \to a$. $\Diamond$

3.7.6. *Remark.* For any bounded sequence (a_n), $b_n \downarrow b$ and $-c_n \downarrow -c$, therefore $b_n - c_n \downarrow b - c$ (2.5.6). Theorem 3.7.5 can thus be reformulated as follows: A sequence (a_n) in $\mathbb{R}$ is convergent if and only if it is bounded and $b_n - c_n \downarrow 0$. This is the promised reformulation of convergence in terms of monotone convergence.

Every bounded sequence (a_n) has a convergent subsequence (3.5.9); in fact, there are subsequences converging to c and to b, and these numbers are, respectively, the smallest and largest possible limits for convergent subsequences:

3.7.7. **Theorem.** *Let* (a_n) *be a bounded sequence in* $\mathbb{R}$ *and let*

$$S = \{x \in \mathbb{R} : a_{n_k} \to x \text{ for some subsequence } (a_{n_k}) \} ;$$

let $c = \liminf a_n$ *and* $b = \limsup a_n$. *Then*

$$\{c, b\} \subset S \subset [c, b] ,$$

thus c *is the smallest element of* S *and* b *is the largest.*

Proof. The first inclusion asserts that each of c and b is the limit of a suitable subsequence of (a_n); for example, let's prove the assertion for b.

Let $\epsilon > 0$. As shown in the proof of 3.7.5, $a_n < b + \epsilon$ ultimately. Also, $a_n > b - \epsilon$ *frequently*; the alternative is that $a_n \leq b - \epsilon$ ultimately (3.2.4), say for $n \geq N$, which would imply that $b_N \leq b - \epsilon < b$, contrary to $b \leq b_N$. Putting these two remarks together, we see that

$$(\forall\, \epsilon > 0) \quad b - \epsilon < a_n < b + \epsilon \quad \text{frequently} .$$

With $\epsilon = 1$, choose n_1 so that

$$b - 1 < a_{n_1} < b + 1 ;$$

with $\epsilon = 1/2$, choose $n_2 > n_1$ so that

$$b - 1/2 < a_{n_2} < b + 1/2 ;$$

continuing in this way, we construct a subsequence (a_{n_k}) such that $|a_{n_k} - b| < 1/k$ for all k; then $a_{n_k} \to b$, so $b \in S$. The proof that $c \in S$ is similar.

To prove the second inclusion, assuming $a_{n_k} \to x$ we have to show that $c \le x \le b$; let's show, for example, that $x \le b$, that is, $x - b \le 0$. Given any $\epsilon > 0$, it suffices to show that $x - b \le \epsilon$ (2.1.2). Indeed, $a_n < b + \epsilon$ for all sufficiently large n, therefore $a_{n_k} < b + \epsilon$ for all sufficiently large k, therefore $x \le b + \epsilon$ by (8) of 3.4.8. The proof that $c \le x$ is similar. $\Diamond$

Exercises

1. Find $\limsup a_n$ and $\liminf a_n$ for each of the following sequences:
(i) $a_n = (-1)^n + 1/n$
(ii) $a_n = (-1)^n(2 + 3/n)$
(iii) $a_n = 1/n + (-1)^n/n^2$
(iv) $a_n = [n + (-1)^n(2n + 1)]/n$

2. Complete the proof of 3.7.7. {That is, show that $c \in S$ and that $c \le x$ for all $x \in S$.}

3. With notations as in 3.7.7, prove that (a_n) is convergent if and only if S is a singleton (that is, $S = \{a\}$ for some $a \in \mathbb{R}$).

4. Observe that Theorem 3.7.7 gives a proof of the Weierstrass-Bolzano theorem independent of the earlier proof (3.5.9).

5. If (a_n) is any bounded sequence, prove that $\liminf a_n = -\limsup (-a_n)$.

6. Let (a_n) and (b_n) be bounded sequences in $\mathbb{R}$. Prove:
(i) $\limsup(a_n + b_n) \le \limsup a_n + \limsup b_n$.
{Hint: Let $c = \limsup(a_n + b_n)$ and let $(a_{n_k} + b_{n_k})$ be a subsequence of $(a_n + b_n)$ converging to c. Passing to a further subsequence, we can suppose that (a_{n_k}) is convergent, say $a_{n_k} \to x$. Then $b_{n_k} \to c - x$, $x \le \limsup a_n$ and $c - x \le \limsup b_n$.}
(ii) $\liminf(a_n + b_n) \ge \liminf a_n + \liminf b_n$.
(iii) If $a_n \ge 0$ and $b_n \ge 0$ for all n, then

$$\limsup(a_n b_n) \le (\limsup a_n)(\limsup b_n).$$

(iv) If $a_n \to a$ then

$$\limsup(a_n + b_n) = a + \limsup b_n,$$
$$\liminf(a_n + b_n) = a + \liminf b_n.$$

(v) Give examples to show that the inequalities in (i)–(iii) may be strict.

7. Let (a_n) be a bounded sequence in $\mathbb{R}$.

(i) Prove that if $b = \limsup a_n$ then b satisfies the following condition:
$(\forall \epsilon > 0)$ $a_n < b + \epsilon$ ultimately and $a_n > b - \epsilon$ frequently.

(ii) Show that the condition in (i) characterizes the limit superior, in the sense that if $b \in \mathbb{R}$ satisfies the condition then necessarily $b = \limsup a_n$.

8. Let (a_n) be a bounded sequence in $\mathbb{R}$.

(i) Prove that if $c = \liminf a_n$ then c satisfies the following condition:
$(\forall \epsilon > 0)$ $a_n > c - \epsilon$ ultimately and $a_n < c + \epsilon$ frequently.

(ii) Show that the condition in (i) characterizes the limit inferior.

{Hint: Exercises 5 and 7.}

Special Subsets of $\mathbb{R}$

The purpose of the chapter is to explore those special subsets of $\mathbb{R}$ (pre-eminently, but not exclusively, intervals) that prove to be most useful in the study of the functions of calculus.

4.1. Intervals

There are 9 kinds of subsets of $\mathbb{R}$ that are called **intervals**. First, there are the 4 kinds of intervals described in 1.3.2:

$$[a, b], \ (a, b), \ [a, b), \ (a, b].$$

These can be visualized as segments on the real line (Figure 4).

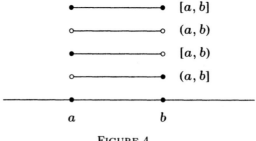

FIGURE 4

{The segments are transposed above the line so as to show the absence or presence of endpoints; for example, the segment with a hole at the left end and a dark bead at the right end represents the interval $(a, b]$.}

Next, for each real number c there are the four 'half-lines'

$$\{x \in \mathbb{R} : x \le c\},$$
$$\{x \in \mathbb{R} : x < c\},$$
$$\{x \in \mathbb{R} : x \ge c\},$$
$$\{x \in \mathbb{R} : x > c\},$$

pictorially represented in Figure 5.

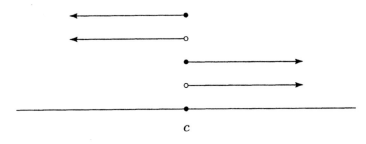

c

FIGURE 5

Finally, $\mathbb{R}$ itself is regarded as an interval (extending indefinitely in both directions):

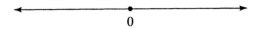

0

The four kinds of bounded intervals have simple representations in terms of their endpoints; the five kinds of unbounded intervals have similar representations, but first we have to invent their endpoints:

4.1.1. *Definition.* For every real number x, we write $x < +\infty$ and $x > -\infty$; in one breath,

$$-\infty < x < +\infty \quad (\forall \, x \in \mathbb{R}).$$

{Think of $+\infty$ (read "plus infinity") as a symbol that stands to the right of every point of the real line, and $-\infty$ ("minus infinity") as a symbol that stands to the left of every point of the line.} To round things out, we write $-\infty < +\infty$.

In effect, a new set $\mathbb{R} \cup \{-\infty, +\infty\}$ has been created, by adjoining to $\mathbb{R}$ two new elements and specifying the order relations between the new elements ($-\infty$ and $+\infty$) and the old ones (the elements of $\mathbb{R}$).

None of this is in the least mysterious: there is a natural correspondence between real numbers x and points P of a semicircle (Figure 6),

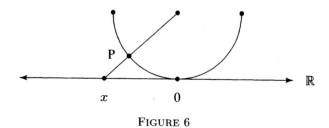

$$x \qquad \cdot \qquad 0$$

FIGURE 6

and the "points at $\pm\infty$" are just what is needed to correspond to the endpoints of the semicircle.

A computationally simpler explanation: the function $f : (-1, 1) \to \mathbb{R}$ defined by $f(x) = x/(1 - |x|)$ is an order-preserving bijection (Figure 7);

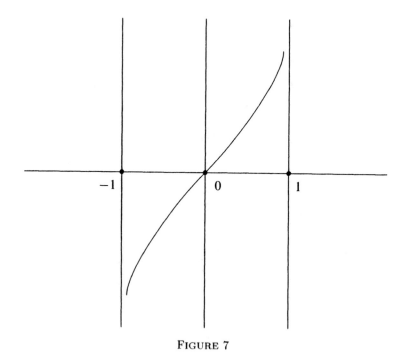

FIGURE 7

it can be extended (in an order-preserving way) to the closed interval $[-1, 1]$ by assigning the values $\pm\infty$ to the endpoints ± 1.

Getting back to the business at hand, the unbounded intervals now have a simple representation:

4.1.2. *Definition.* For any real number c, we write

$$[c, +\infty) = \{x \in \mathbb{R} : c \le x < +\infty\} = \{x \in \mathbb{R} : x \ge c\},$$
$$(c, +\infty) = \{x \in \mathbb{R} : c < x < +\infty\} = \{x \in \mathbb{R} : x > c\},$$
$$(-\infty, c] = \{x \in \mathbb{R} : -\infty < x \le c\} = \{x \in \mathbb{R} : x \le c\},$$
$$(-\infty, c) = \{x \in \mathbb{R} : -\infty < x < c\} = \{x \in \mathbb{R} : x < c\},$$
$$(-\infty, +\infty) = \{x \in \mathbb{R} : -\infty < x < +\infty\} = \mathbb{R}.$$

{Note that when $+\infty$ or $-\infty$ (neither of which is a real number) is used as an 'endpoint' of an interval of $\mathbb{R}$, it is always *absent* from the interval, therefore it is always adjacent to a parenthesis (never a square bracket).}

To summarize it all for convenient reference:

4.1.3. *Definition.* An **interval** of $\mathbb{R}$ is a subset of $\mathbb{R}$ of one of the following 9 types:

$$[a, b], \ (a, b), \ [a, b), \ (a, b],$$

$$[c, +\infty), \ (c, +\infty), \ (-\infty, c], \ (-\infty, c), \ (-\infty, +\infty) = \mathbb{R},$$

where a, b, c are real numbers and $a \le b$.

In particular, the empty set $\varnothing = (a, a) = [a, a) = (a, a]$ and singletons $\{a\} = [a, a]$ qualify as intervals, albeit 'degenerate'.

The intervals of $\mathbb{R}$, of all 9 types, are characterized by a single property (called *convexity*):

4.1.4. **Theorem.** *Let* A *be a nonempty subset of* $\mathbb{R}$. *The following conditions are equivalent:*

(a) A *is an interval;*

(b) *for every pair of points in* A, *the segment joining them is contained in* A; *that is,*

$$x, y \in A, \ x \le y \ \Rightarrow \ [x, y] \subset A.$$

Proof. The implication (a) $\Rightarrow$ (b) is obvious.

(b) $\Rightarrow$ (a): There are four cases, according as A is bounded above (or not) and bounded below (or not).

case 1: A bounded below, but not bounded above.

Let $a = \inf A$. We will show that $A = (a, +\infty)$ or $A = [a, +\infty)$; for this, it suffices to show that

$$(a, +\infty) \subset A \subset [a, +\infty).$$

The second inclusion is immediate from the definition of a. Assuming $r \in (a, +\infty)$, we have to show that $r \in A$. Since $r > a = \text{GLB} A$, there exists $x \in A$ such that $a \le x < r$. {Reason: a is the *greatest* lower bound of A, and $r > a$, so r can't be a lower bound, etc.} On the other hand, since A is not bounded above, there exists $y \in A$ such that $y > r$. Thus $x < r < y$ with $x, y \in A$, so $r \in [x, y] \subset A$ by the hypothesis (b).

The remaining three cases are easily proved using appropriate variations on the preceding arguments; they are left as an exercise, with the following sketch as hint.

case 2: A bounded above, but not below.
With $b = \sup A$, one argues similarly that

$$(-\infty, b) \subset A \subset (-\infty, b].$$

case 3: A bounded (above and below).
With $a = \inf A$ and $b = \sup A$, one argues that

$$(a, b) \subset A \subset [a, b],$$

so A is one of the four kinds of bounded interval.

case 4: A neither bounded above nor below.
For every $r \in \mathbb{R}$ there exist $x, y \in A$ with $x < r < y$, so $r \in [x, y] \subset A$; thus $A = \mathbb{R}$. ◊

4.1.5. Corollary. *If $\mathcal{S}$ is any set of intervals and $J = \bigcap \mathcal{S}$ is their intersection, then J is an interval (possibly empty).*

Proof. By definition, J is the set of all real numbers common to all of the intervals belonging to $\mathcal{S}$, that is,

$$J = \{r \in \mathbb{R} : r \in I \text{ for every } I \in \mathcal{S}\}.$$

Assuming J nonempty, it will suffice to verify condition (b) of the theorem. Suppose $x, y \in J$, $x \leq y$. Then $x, y \in I$ for every $I \in \mathcal{S}$; by the theorem, $[x, y] \subset I$ for all $I \in \mathcal{S}$, consequently $[x, y] \subset J$. ◊

Exercises

1. Complete the proof of 4.1.4 (especially cases 2 and 3).

2. (i) If I and J are intervals in $\mathbb{R}$ such that $I \cap J \neq \emptyset$, prove that $I \cup J$ is an interval.

(ii) True or false (explain): If I and J are nonempty intervals such that $I \cup J$ is an interval, then $I \cap J \neq \emptyset$.

(iii) Same question as (ii), assuming I and J are open intervals.

(iv) Same question as (ii), assuming I and J are closed intervals.

3. Let $\mathcal{S}$ be a set of intervals in $\mathbb{R}$ such that, for every pair of intervals I, J in $\mathcal{S}$, there exists $K \in \mathcal{S}$ such that $I \cup J \subset K$. Prove that the union $\bigcup \mathcal{S}$ of all the intervals in $\mathcal{S}$ is an interval.

4. Let S be a nonempty subset of $\mathbb{R}$, let $\mathcal{S}$ be the set of all closed intervals whose endpoints are in S, and let $A = \bigcup \mathcal{S}$. Prove that A is an interval.

5. Let A be a subset of $\mathbb{R}$ with the following property: for every $x \in A$ there exists $y \in A$ such that $x < y$ and $[x, y] \subset A$. True or false (explain): A is an interval.

6. How would you extend the operation $x \mapsto -x$ from $\mathbb{R}$ to $\mathbb{R} \cup \{-\infty, +\infty\}$? Why?

7. True or false (explain): $\bigcap_{n=1}^{\infty} [n, +\infty) = \varnothing$.

8. Let I be an interval, (x_n) a sequence in I that converges to a point x of I. Prove that there exists a closed interval $J \subset I$ that contains x and every x_n.

{Hint: Translating by $-x$, one can suppose that $x_n \to 0$. Passing to subsequences, reduce to the case that $x_n \geq 0$ for all n. It then suffices to show that the sequence has a largest term, and for this it is enough to find an index N such that $x_n \leq x_N$ ultimately. The alternative is that $(\forall N)\, x_n > x_N$ frequently, whence a subsequence (x_{n_k}) with $x_{n_1} < x_{n_2} < x_{n_3} < \ldots$; but this is ruled out by $x_{n_k} \to 0$.}

9. (i) Every interval is the union of an increasing sequence of closed intervals.

(ii) The union of an increasing sequence of intervals is an interval.

4.2. Closed Sets

The most important subsets of $\mathbb{R}$ for calculus are the intervals. There are differences among intervals, some important, others not (depending, in part, on the context). For example, the difference between $(0, 1)$ and $(0, 5)$ is only a matter of scale; the inequalities defining them are qualitatively the same. The intervals $(0, 1)$ and $(0, +\infty)$ are different in kind, one being bounded and the other not; even so, the difference is somewhat muted by the examples following 4.1.1.

By contrast, the intervals $I = (0, 1)$ and $J = [0, 1]$ prove to have dramatically different properties (§4.5); the crux of the matter is that the endpoints $0, 1$ of I can be approximated as closely as we like by points of I but they are not themselves points of I. More precisely, the endpoints of I, though not in I, are limits of convergent sequences whose terms are in I. On the other hand, if a convergent sequence has its terms in J then its limit must also be in J, by (8) of 3.4.8. The latter property is meaningful for sets that are not necessarily intervals:[1]

---

[1]This is the point at which our surreptitious study of the 'topology' of $\mathbb{R}$ begins (cf. the remarks in the Preface).

4.2.1. *Definition.* A set A of real numbers is said to be a **closed** subset
of $\mathbb{R}$ (or to be a **closed set** in $\mathbb{R}$) if, whenever a convergent sequence has
all of its terms in A , the limit of the sequence must also be in A . That is,
if $x_n \to x$ and $x_n \in A$ for all n, then necessarily $x \in A$. {Suggestively:
You can't escape from a closed set by means of a convergent sequence!} In
symbols,

$$\left.\begin{array}{c} x_n \in A \ (\forall \ n) \\ x \in \mathbb{R} \\ x_n \to x \end{array}\right\} \Rightarrow x \in A.$$

Convention: The empty subset $\varnothing$ of $\mathbb{R}$ is closed.[2]

4.2.2. *Example.* $\mathbb{R}$ is a closed subset of $\mathbb{R}$ (nowhere else for the limit
to go!).

4.2.3. *Example.* Every singleton $\{a\}$ $(a \in \mathbb{R})$ is closed (the constant
sequence $x_n = a$ converges to a).

4.2.4. *Example.* For every real number c, the intervals $[c, +\infty)$ and
$(-\infty, c]$ are closed sets. For example, if $x_n \to x$ and $x_n \geq c$ for all n,
then $x \geq c$ by (8) of 3.4.8. {Caution: These are closed sets and they are
intervals, but they are not closed intervals; the latter term is reserved for
intervals $[a, b]$.}

Before enlarging the list of examples, let's prove some useful general
properties of closed sets:

4.2.5. **Lemma.** *If* A *is a closed subset of* $\mathbb{R}$, $x_n \to x$ *in* $\mathbb{R}$, *and*
$x_n \in A$ *frequently, then* $x \in A$.

Proof. By assumption, there is a subsequence (x_{n_k}) with $x_{n_k} \in A$ for
all k; since $x_{n_k} \to x$, $x \in A$ by the definition of a closed set. $\Diamond$

4.2.6. **Theorem.** (i) $\varnothing$ *and* $\mathbb{R}$ *are closed sets in* $\mathbb{R}$.
(ii) *If* A *and* B *are closed sets in* $\mathbb{R}$, *then so is their union* $A \cup B$.
(iii) *If* $\mathcal{S}$ *is any set of closed sets in* $\mathbb{R}$, *then their intersection* $\bigcap \mathcal{S}$ *is
also a closed set.*

Proof. (i) Already noted in 4.2.1 and 4.2.2.
(ii) Suppose $x_n \in A \cup B$ for all n and $x_n \to x$. If $x_n \in A$ frequently
then $x \in A$ by the lemma; the alternative is that $x_n \in B$ ultimately, in
which case $x \in B$, again by the lemma. Either way, $x \in A \cup B$.
(iii) Let $B = \bigcap \mathcal{S} = \{x \in \mathbb{R} : x \in A$ for all $A \in \mathcal{S}\}$. Suppose $x_n \to x$
and $x_n \in B$ for all n. For each $A \in \mathcal{S}$, $x_n \in A$ for all n, therefore

---

[2]Rationale: For A $= \varnothing$ the statement on the left side of the implication
is never true, so the implication is true by 'vacuous implication.' Think of the
implication as a pledge: as soon as someone presents us with an $x \in \mathbb{R}$ and
a sequence (x_n) satisfying the left side of the implication, we stand ready to
verify that x belongs to A ; we're waiting...and, meanwhile, the *pledge* is
valid.

$x \in A$ (because A is closed); thus $x \in A$ for all $A \in \mathcal{S}$, therefore $x \in \bigcap \mathcal{S} = B$. $\Diamond$

4.2.7. Example. Every closed interval

$$[a, b] = (-\infty, b] \cap [a, +\infty)$$

is a closed set (in view of 4.2.4, it is the intersection of two closed sets).

4.2.8. Example. If $A_1, \ldots, A_r$ is a finite list of closed sets in $\mathbb{R}$, then their union $A_1 \cup \ldots \cup A_r$ is also a closed set. {Induction on (ii) of 4.2.6.}

4.2.9. Example. Every finite subset $A = \{a_1, \ldots, a_r\}$ of $\mathbb{R}$ is a closed set; for, $A = \{a_1\} \cup \ldots \cup \{a_r\}$ is closed by 4.2.3 and 4.2.8.

If A is a closed set, we know where the limits of its convergent sequences are (they are in A). On the other hand, the set $A = (0, 1]$ contains a convergent sequence—for example $x_n = 1/n$—whose limit is not in A. For an arbitrary subset A of $\mathbb{R}$ we may contemplate the set $\overline{A}$ of all real numbers that are limits of sequences whose terms are in A; regardless of the status of A, $\overline{A}$ is always a closed set:

4.2.10. Theorem. *Let A be any subset of $\mathbb{R}$ and let*

$$\overline{A} = \{x \in \mathbb{R} : a_n \to x \text{ for some sequence } (a_n) \text{ in } A\}.$$

Then $\overline{A}$ is the smallest closed set containing A, that is,

(1) $\overline{A} \supset A$,

(2) $\overline{A}$ *is a closed set,*

(3) *if B is a closed set with $B \supset A$ then $B \supset \overline{A}$.*

Moreover,

(4) A *is closed* $\Leftrightarrow$ $\overline{A} = A$;

(5) $\overline{A}$ *is the set of all real numbers that can be approximated as closely as we like by elements of A, that is,*

$$x \in \overline{A} \Leftrightarrow (\forall \, \epsilon > 0) \ \exists \, a \in A \ni |x - a| < \epsilon.$$

Proof. (1) If $a \in A$ let $a_n = a$ for all n; then $a_n \to a$, so $a \in \overline{A}$.

(3) Assuming B is a closed set with $A \subset B$, we have to show that $\overline{A} \subset B$. Let $x \in \overline{A}$, say $a_n \to x$ with $a_n \in A$ for all n; then $a_n \in B$ for all n (because $A \subset B$), therefore $x \in B$ (because B is closed).

(4) To say that A is closed means (4.2.1) that $\overline{A} \subset A$; since $\overline{A} \supset A$ automatically, the condition $\overline{A} \subset A$ is equivalent to $\overline{A} = A$.

(5), $\Leftarrow$: For each positive integer n let $\epsilon = 1/n$ and choose $a_n \in A$ such that $|x - a_n| < 1/n$. Then $a_n \to x$, so $x \in \overline{A}$.

$\Rightarrow$: Let $x \in \overline{A}$, say $a_n \to x$ with $a_n \in A$ for all n. If $\epsilon > 0$ then $|x - a_n| < \epsilon$ for some n (in fact, ultimately!).

(2) Assuming $x_n \to x$ with $x_n \in \overline{A}$ for all n, we have to show that $x \in \overline{A}$. Let's apply the criterion of (5): If $\epsilon > 0$, choose n so that $|x - x_n| < \epsilon/2$; for this n, choose $a \in A$ so that $|x_n - a| < \epsilon/2$. Then $|x - a| < \epsilon/2 + \epsilon/2$ by the triangle inequality. $\Diamond$

4.2.11. *Definition.* With notations as in the preceding theorem, $\overline{A}$ is called the **closure** of A in $\mathbb{R}$. {Alternate terminology: The points of $\overline{A}$ are said to be **adherent** to A, and $\overline{A}$ is called the **adherence** of A.}

Exercises

1. (i) Prove: For each $c \in \mathbb{R}$, neither of the intervals $(c, +\infty)$ and $(-\infty, c)$ is a closed set.

(ii) Prove: If $a < b$ then none of the intervals (a, b), $(a, b]$, $[a, b)$ is a closed set.

(iii) Find the closure in $\mathbb{R}$ of each of the intervals in (i) and (ii).

2. Prove that $\mathbb{Z}$ is a closed subset of $\mathbb{R}$, but $\mathbb{Q}$ is not.

3. If A is a bounded nonempty subset of $\mathbb{R}$, prove that $\sup A \in \overline{A}$ and $\inf A \in \overline{A}$.

4. Let $\mathbb{I} = \mathbb{R} - \mathbb{Q}$ be the set of all irrational numbers. Prove that $\overline{\mathbb{I}} = \overline{\mathbb{Q}} = \mathbb{R}$.

5. If A is a closed subset of $\mathbb{R}$, prove that each of the following sets B is also closed:

(i) $B = \{-x : x \in A\}$;

(ii) $B = \{|x| : x \in A\}$;

(iii) $B = \{x^2 : x \in A\}$.

{Hint for (ii) and (iii): Weierstrass-Bolzano theorem.}

6. (i) If A and B are any two subsets of $\mathbb{R}$, prove that

$$\overline{A \cup B} = \overline{A} \cup \overline{B}.$$

(ii) Disprove (with a counterexample): $\overline{A \cap B} = \overline{A} \cap \overline{B}$. {Hint: Exercise 4.}

(iii) Disprove (with a counterexample):

$$\overline{\bigcup_{n=1}^{\infty} A_n} = \bigcup_{n=1}^{\infty} \overline{A_n}.$$

{Hint: List the rational numbers in a sequence (r_n) and let $A_n = \{r_n\}$.}

7. Let (a_n) be a sequence in $\mathbb{R}$ and let

$$S = \{x \in \mathbb{R} : a_{n_k} \to x \text{ for some subsequence } (a_{n_k})\}$$

(conceivably $S = \varnothing$). Prove:
  (i) $S = \{x \in \mathbb{R} : (\forall \, \epsilon > 0) \ |a_n - x| \leq \epsilon \ \text{frequently}\}$;
  (ii) S is a closed subset of $\mathbb{R}$.
  {Hint for (i): Proof of 3.7.7. Hint for (ii): Proof of (2) of 4.2.10.}

  8. Let (X, d) be a metric space (§3.5, Exercise 8) and let A be a subset of X. Imitating 4.2.1, call A a *closed* subset of X if, whenever $x_n \in A$ ($n \in \mathbb{P}$) and $x_n \to x$ (in the sense of §3.4, Exercise 12), necessarily $x \in A$. Guided by 4.2.11, define the closure $\overline{A}$ of a subset A of X.
  (i) Prove the metric space analogues of 4.2.2, 4.2.3, 4.2.5, 4.2.6, 4.2.8–4.2.10.
  (ii) Let c be a fixed point of X. For any $r > 0$ consider the sets

$$\{x \in X : d(x, c) \leq r\}, \quad \{x \in X : d(x, c) \geq r\}$$

and prove the analogues of 4.2.4 and 4.2.7.

  9. For a subset A of $\mathbb{R}$, the following conditions are equivalent: (a) A is closed; (b) if $x_n \to x$ and $x_n \in A$ frequently, then $x \in A$; (c) if $x_n \to x$ and $x_n \in A$ ultimately, then $x \in A$.

4.3. Open Sets, Neighborhoods

  According to Theorem 4.2.10, the meaning of $x \in \overline{A}$ is that for every $\epsilon > 0$ the interval $(x - \epsilon, x + \epsilon)$ intersects A, that is,

$$(\forall \, \epsilon > 0) \ (x - \epsilon, x + \epsilon) \cap A \neq \varnothing.$$

The meaning of $x \notin \overline{A}$ is the negation of the preceding condition: there exists an $\epsilon > 0$ for which the interval $(x - \epsilon, x + \epsilon)$ is disjoint from A, that is,

$$\exists \, \epsilon > 0 \ni (x - \epsilon, x + \epsilon) \cap A = \varnothing,$$

in other words,

$$\exists \, \epsilon > 0 \ni (x - \epsilon, x + \epsilon) \subset \complement A,$$

where $\complement A = \mathbb{R} - A$ is the complement of A; so to speak, not only does x belong to $\complement A$, but there is a little buffer zone about x that remains in $\complement A$—informally, all points 'sufficiently close to x' are in $\complement A$. There is a technical term for this idea:

  4.3.1. *Definition.* A point $x \in \mathbb{R}$ is said to be **interior** to a subset A of $\mathbb{R}$ if there exists an $r > 0$ such that $(x - r, x + r) \subset A$, that is, such that

$$|y - x| < r \ \Rightarrow \ y \in A.$$

If x is interior to A, one also says that A is a **neighborhood** of x .[1]
The set of all interior points of A (there may not be any!) is called the
interior of A, denoted $A°$:

$$A° = \{x \in \mathbb{R} : x \text{ is interior to } A\};$$

thus, $A°$ is the set of all points of A of which A is a neighborhood.

4.3.2. *Examples.* $\mathbb{Q}$ has no interior points (i.e., it has empty interior), because every open interval contains an irrational number (§2.4, Exercise 3). Assuming $a < b$, the point a belongs to $[a, b)$ but not to its
interior; the interior of $[a, b]$ is (a, b).

The discussion preceding 4.3.1 can be summarized as follows:

4.3.3. Theorem. *If A is any subset of $\mathbb{R}$, then*

$$x \notin \overline{A} \Leftrightarrow x \in (\complement A)°;$$

thus $\complement \overline{A} = (\complement A)°$.

In words, the complement of the closure (of a set) is the interior of the
complement (of the set). The formula in 4.3.3 can be written

$$\overline{A} = \complement(\complement A)°;$$

in effect, the passage from A to its closure $\overline{A}$ is the composite of three
operations: take complement, then take interior, then take complement
again. Applying the formula to $\complement A$ in place of A, we have

$$A° = \complement \overline{\complement A},$$

thus the interior of A is the complement of a closed set (we'll have more
to say about this shortly).

In general, $A \subset \overline{A}$; equality is a special event (A closed). In general,
$A° \subset A$; again, equality is a special event:

4.3.4. *Definition.* A subset A of $\mathbb{R}$ is called an **open set** if every point
of A is an interior point, that is,

$$(\forall x \in A) \; \exists \epsilon > 0 \ni (x - \epsilon, x + \epsilon) \subset A.$$

(Equivalently, A is a neighborhood of each of its points.)

Intuitively, for every point of an open set A, there is a buffer zone about
the point—whose size may depend on the point—that is also contained
in A.

---

[1] The relation 'A is a neighborhood of x' is just an alternate way of saying
that 'x is interior to A'. Analogy: $b > a$ is just another way of saying $a < b$.

To say that A is open means that $A \subset A^\circ$; since $A \supset A^\circ$ automatically, an equivalent condition is that $A = A^\circ$, in other words, A is equal to its interior. The analogy with closed sets is no accident:

4.3.5. Theorem.[2] *For a subset* A *of* $\mathbb{R}$,
(i) A *is open* $\Leftrightarrow$ $\complement A$ *is closed*;
(ii) A *is closed* $\Leftrightarrow$ $\complement A$ *is open*.

Proof. (i) In general, $A^\circ = \complement \overline{\complement A}$, so the following conditions are equivalent: A open, $A = A^\circ$, $A = \complement \overline{\complement A}$, $\complement A = \overline{\complement A}$, $\complement A$ closed.
(ii) Apply (i) with A replaced by $\complement A$. ◊

4.3.6. Corollary. *For every subset* A *of* $\mathbb{R}$, A° *is the largest open subset of* A.

Proof. By the remarks following 4.3.3, A° is the complement of a closed set, so it is open. On the other hand, if U is any open set with $U \subset A$, then $U \subset A^\circ$; for, if $x \in U$ then x is interior to U, so it is obviously interior to A as well. Summarizing, A° is is an open subset of A, and it contains all others. ◊

4.3.7. Corollary. *Let* $A \subset \mathbb{R}$, $x \in \mathbb{R}$. *The following conditions are equivalent*:
(a) A *is a neighborhood of* x;
(b) *there exists an open set* U *such that* $x \in U \subset A$.

Proof. (a) $\Rightarrow$ (b): $U = A^\circ$ fills the bill (4.3.6).
(b) $\Rightarrow$ (a): Since $x \in U$ and U is open, U is a neighborhood of x, therefore so is its superset A (clear from 4.3.1). ◊

The next theorem is the open set analogue of Theorem 4.2.6:

4.3.8. Theorem. (i) $\varnothing$ *and* $\mathbb{R}$ *are open sets in* $\mathbb{R}$.
(ii) *If* A *and* B *are open sets in* $\mathbb{R}$, *then so is their intersection* $A \cap B$.
(iii) *If* $\mathcal{S}$ *is any set of open sets in* $\mathbb{R}$, *then their union* $\bigcup \mathcal{S}$ *is also an open set.*

Proof. (i) $\varnothing = \complement \mathbb{R}$ and $\mathbb{R} = \complement \varnothing$ are the complements of closed sets (4.2.6), hence are open (4.3.5). The proofs of (ii) and (iii) could also be based on 4.2.6 and 4.3.5, but we learn more by going back to first principles:
(ii) Assuming $x \in A \cap B$ we have to show that x is interior to $A \cap B$. Since $x \in A$ and A is open, there is an $r > 0$ with $(x - r, x + r) \subset A$. Similarly, there is an $s > 0$ with $(x - s, x + s) \subset B$. If t is the smaller of r and s, then
$$(x - t, x + t) \subset A \cap B,$$
so x is interior to $A \cap B$.

---

[2]Caution: In life, "open" generally means "not closed"; but in $\mathbb{R}$, a subset may be neither closed nor open, for example the interval $[0, 1)$.

(iii) Let $B = \bigcup \mathcal{S}$. If $x \in B$ then $x \in A$ for some $A \in \mathcal{S}$. By assumption, A is open, so there is an $r > 0$ with $(x - r, x + r) \subset A$; since $A \subset B$, it follows that $(x - r, x + r) \subset B$, thus x is interior to B. We have shown that every point of B is an interior point, consequently B is open. $\diamondsuit$

4.3.9. The proof of (ii) of 4.3.8 shows that if A and B are neighborhoods of x then so is $A \cap B$.

Exercises

1. Show that $x \in \overline{A}$ if and only if every neighborhood of x intersects A.

2. (i) Show that $(A \cap B)^\circ = A^\circ \cap B^\circ$ for all subsets A and B of $\mathbb{R}$.
(ii) True or false (explain): $(A \cup B)^\circ = A^\circ \cup B^\circ$ for all subsets A and B of $\mathbb{R}$.

3. If A is a subset of $\mathbb{R}$ and U is an open set in $\mathbb{R}$, prove that

$$\overline{A \cap U} \supset \overline{A} \cap U.$$

{Hint: If $x \in \overline{A} \cap U$ and V is a neighborhood of x, then $U \cap V$ is also a neighborhood of x.}

4. Let A be a subset of $\mathbb{R}$ and, for each positive integer n, let

$$U_n = \{x \in \mathbb{R} : |x - a| < 1/n \text{ for some } a \in A\}.$$

Prove:
(i) U_n is an open set. {Hint: Exhibit it as the union of a set of open intervals.}
(ii) $\overline{A} = \bigcap_{n=1}^{\infty} U_n$.
(iii) Every closed set in $\mathbb{R}$ is the intersection of a sequence of open sets.

5. Prove that every open set in $\mathbb{R}$ is the union of a sequence of closed sets. {Hint: 4.3.5.}

6. (i) Find a sequence of open intervals whose intersection is not an open set.
(ii) Find a sequence of closed intervals whose union is not a closed set.

7. A subset A of $\mathbb{R}$ is said to be *dense* in $\mathbb{R}$ if its closure is $\mathbb{R}$, that is, $\overline{A} = \mathbb{R}$. Prove that the following conditions are equivalent:
(a) A is dense in $\mathbb{R}$;
(b) for every nonempty open set U in $\mathbb{R}$, $U \cap A \neq \varnothing$; that is, A intersects every nonempty open set in $\mathbb{R}$.

8. Prove that A is a dense subset of $\mathbb{R}$ (Exercise 7) if and only $\mathbb{R} - A$ has empty interior.

9. (i) Prove that $\mathbb{Q}$ and $\mathbb{R} - \mathbb{Q}$ are dense subsets of $\mathbb{R}$ (cf. Exercise 7).

(ii) Exhibit a subset A of $\mathbb{R}$, with $A \neq \mathbb{R}$, such that A is dense but $\mathbb{R} - A$ is not.

(iii) If U and V are dense open sets in $\mathbb{R}$, prove that $U \cap V$ is also dense.

(iv) Let (x_n) be any sequence in $\mathbb{R}$ and let $A = \mathbb{R} - \{x_n : n \in \mathbb{P}\}$ be the remaining points of $\mathbb{R}$. Prove that A is dense in $\mathbb{R}$. {Hint: $\complement A$ contains no nondegenerate closed interval (§2.6, Exercise 4); cf. Exercise 8.}

10. Let (X, d) be a metric space (cf. §3.5, Exercise 8). If $c \in X$ and $r > 0$, the set
$$U_r(c) = \{x \in X : d(x, c) < r\}$$
is called the *open ball* in X with *center* c and *radius* r, and the set
$$B_r(c) = \{x \in X : d(x, c) \leq r\}$$
is called the *closed ball* with center c and radius r.

Let $A \subset X$, $x \in X$. In analogy with 4.3.1, x is said to be *interior* to A (and A is called a *neighborhood* of x) if there exists an $r > 0$ such that $U_r(x) \subset A$; the set of all interior points of A is called the *interior* of A, denoted A°; A is said to be *open* if every point of A is an interior point (cf. 4.3.4).

(i) Show that if A and B are neighborhoods of a point x, then so is $A \cap B$.

(ii) Every closed ball is a closed set (§4.2, Exercise 8). Prove that every open ball is an open set.

(iii) Prove the metric space analogues of 4.3.3 and 4.3.5–4.3.8.

11. In a metric space (cf. Exercise 10) the following conditions on a point x and a subset A are equivalent:

(a) x is interior to A;

(b) if $x_n \to x$ then $x_n \in A$ ultimately.

{Hint: (a) $\Rightarrow$ (b) is easy. If (a) is false then, for each $n \in \mathbb{P}$, the open ball $U_{1/n}(x)$ must contain a point x_n that is not in A; infer that (b) is false.}

12. Let $\mathcal{C}$ be a set of nonempty open sets in $\mathbb{R}$ such that for $U, V \in \mathcal{C}$, either $U = V$ or $U \cap V = \emptyset$ (the sets in $\mathcal{C}$ are then said to be 'pairwise disjoint'). Prove: Either $\mathcal{C}$ is finite, or the sets in $\mathcal{C}$ can be listed in a sequence (U_n).

{Hint: Enumerate the rational numbers in a sequence (r_n); in each $U \in \mathcal{C}$ choose a rational number.}

13. Let U be a nonempty open set in $\mathbb{R}$. Prove that U is the union of a set $\mathcal{C}$ of pairwise disjoint intervals that are open sets. {Hint: For each

$x \in U$, write I_x for the set of all $y \in U$ such that the closed interval with endpoints x and y is contained in U; by 4.1.4, I_x is an interval. Let C be the set of intervals I_x.}

14. If A is a nonempty closed subset of $\mathbb{R}$ such that $A \neq \mathbb{R}$, then $\complement A$ is not closed; thus, the only subsets of $\mathbb{R}$ that are both closed and open are $\varnothing$ and $\mathbb{R}$.
{Hint: Choose $a \in A$, $b \in \complement A$. If, for example, $a < b$, let $c = \sup(A \cap [a,b])$ and argue that $c \in A$ (cf. §4.2, Exercise 3) whereas $(c,b] \subset \complement A$.}

15. If p is a nonzero polynomial with real coefficients, then $\{x \in \mathbb{R} : p(x) \neq 0\}$ is an open set.

16. Let I be an interval with endpoints $a < b$. If U is an open set in $\mathbb{R}$ such that $U \cap I \neq \varnothing$, then $U \cap (a,b) \neq \varnothing$.

17. A subset A of $\mathbb{R}$ is said to be *connected* if it has the following property: If U and V are open sets in $\mathbb{R}$ such that $A \subset U \cup V$, $U \cap A \neq \varnothing$ and $V \cap A \neq \varnothing$, then $U \cap V \cap A \neq \varnothing$; in other words, if U and V are open sets such that

$$A = (U \cap A) \cup (V \cap A) \quad \text{and} \quad (U \cap A) \cap (V \cap A) = \varnothing,$$

then either $U \cap A = \varnothing$ or $V \cap A = \varnothing$. Examples: $\varnothing$ is connected (trivial); $\mathbb{R}$ is connected (Exercise 14).
The following propositions culminate in a proof that a subset of $\mathbb{R}$ is connected if and only if it is an interval.
(i) Every connected set A is an interval. {Hint: If $a, b \in A$ and $a < c < b$, it suffices to show that $c \in A$ (4.1.4). If $c \notin A$ consider the sets $U = (-\infty, c)$ and $V = (c, +\infty)$.}
(ii) Every closed interval $[a,b]$ is connected. {Hint: Let U and V be open sets such that

$$[a,b] = (U \cap [a,b]) \cup (V \cap [a,b]),$$

$$(U \cap [a,b]) \cap (V \cap [a,b]) = \varnothing.$$

Let $A = U \cap [a,b]$, $B = V \cap [a,b]$ and assume to the contrary that A and B are both nonempty. Note that A and B are closed sets in $\mathbb{R}$; for example, if $x_n \in A$ and $x_n \to x$, the alternative to $x \in A$ is $x \in B$, whence $x_n \in V$ ultimately, contrary to $A \cap B = \varnothing$. If $c = \sup A$, then $c \in A$ (§4.2, Exercise 3). Necessarily $c = b$; for, if $c < b$ then there exists an $\epsilon > 0$ such that $[c - \epsilon, c + \epsilon] \subset U$ and $c + \epsilon < b$, which leads to the absurdity $c + \epsilon \in A$. Thus $b = \sup A$. Similarly, $b = \sup B$, so $b \in A \cap B$, a contradiction.}
(iii) If $A = \bigcup A_n$, where (A_n) is a sequence of connected sets such that $A_1 \subset A_2 \subset A_3 \subset \ldots$, then A is connected. {Hint: Let U and V be open sets with

$$A = (U \cap A) \cup (V \cap A) \quad \text{and} \quad U \cap V \cap A = \varnothing.$$

For each n, $A_n = (U \cap A_n) \cup (V \cap A_n)$ with the terms of the union disjoint, so either $U \cap A_n = \varnothing$ or $V \cap A_n = \varnothing$. If $U \cap A_n = \varnothing$ frequently, then $U \cap A_n = \varnothing$ for all n, whence $U \cap A = \varnothing$.}

(iv) Every interval in $\mathbb{R}$ is connected. {Hint: Every interval is the union of an increasing sequence of closed intervals.}

18. There exists a sequence of open intervals (a_n, b_n) such that every open set in $\mathbb{R}$ is the union of certain of the (a_n, b_n). {Hint: Think rational endpoints.}

19. For a subset A of $\mathbb{R}$, the following conditions are equivalent: (a) A is open; (b) if $x_n \to x$ and $x \in A$, then $x_n \in A$ ultimately; (c) if $x_n \to x$ and $x \in A$, then $x_n \in A$ frequently. (Cf. §4.2, Exercise 9.)

4.4. Finite and Infinite Sets

The subject of the next section is a famous theorem about "finite coverings"; to appreciate it, we should pause to reflect on the mathematical usage of the terms "finite" and "infinite".[1]

Intuitively, a set is finite if, for some positive integer r, the elements of the set can be labeled with the integers from 1 to r. More formally:

4.4.1. _Definition._ A nonempty set A is said to be **finite** if there exist a positive integer r and a surjection $\{1, \ldots, r\} \to A$. Convention: The empty set $\varnothing$ is finite. A set is said to be **infinite** if it is not finite.

Finite sets are often presented in the following form. If $\sigma : \{1, \ldots, r\} \to A$ is a surjection and one writes $x_i = \sigma(i)$ for $i = 1, \ldots, r$, then

$$A = \sigma(\{1, \ldots, r\}) = \{x_1, \ldots, x_r\}.$$

We also say that $x_1, \ldots, x_r$ is a _finite list_ of elements.[2]

4.4.2. _Example._ For each positive integer r, the set $\{1, \ldots, r\}$ is finite. {Proof: The identity mapping $\{1, \ldots, r\} \to \{1, \ldots, r\}$ is a surjection.}

4.4.3. _Example._ The set $\mathbb{P}$ of all positive integers is infinite. {Proof: We have to show that there does not exist a surjection $\{1, \ldots, r\} \to \mathbb{P}$ for any r. In other words, assuming $r \in \mathbb{P}$ and $\varphi : \{1, \ldots, r\} \to \mathbb{P}$, we have to show that φ is not surjective. Let $n = 1 + \varphi(1) + \ldots + \varphi(r)$; then $\varphi(i) < n$ for all $i = 1, \ldots, r$, therefore n is not in the range of φ.}

---

[1] Suggestion for a first reading: Skim through the definitions and the statements of the theorems. If the statements look plausible and the proofs feel superfluous, it is not a sin to skip over the details. (_Someday_—not necessarily today—you'll want to worry about such matters.) The underlying challenge: when we use the words, do we know what we're talking about?

[2] The word "list" often also connotes the particular order in which the elements are written down; the list x_2, x_1, x_3 would then be distinguished from x_1, x_2, x_3, although the _sets_ $\{x_2, x_1, x_3\}$ and $\{x_1, x_2, x_3\}$ are equal.

4.4.4. Theorem. *If* $f : X \to Y$ *is any function and* A *is a finite subset of* X, *then* $f(A)$ *is a finite subset of* Y.

Proof. If $\sigma : \{1, \ldots, r\} \to A$ is surjective, then $i \mapsto f(\sigma(i))$ is a surjection $\{1, \ldots, r\} \to f(A)$. ◊

4.4.5. Theorem. *If* $A_1, \ldots, A_m$ *is a finite list of finite subsets of a set, then* $A_1 \cup \ldots \cup A_m$ *is also finite.*

Proof. For each $j = 1, \ldots, m$ there is a positive integer r_j and a surjection $\sigma_j : \{1, \ldots, r_j\} \to A_j$. Let $r = r_1 + \ldots + r_m$; we will construct a surjection

$$\sigma : \{1, \ldots, r\} \to A_1 \cup \ldots \cup A_m$$

in the 'obvious' way. (But it still takes some concentration to put it down on paper!)

The elements of $\{1, \ldots, r\}$ can be organized, in ascending order, as a union of m subsets:

$$\{1, \ldots, r\} = \{1, \ldots, r_1\} \cup \{r_1 + 1, \ldots, r_1 + r_2\} \cup \ldots$$
$$\cup \{r_1 + \ldots + r_{m-1} + 1, \ldots, r_1 + \ldots + r_{m-1} + r_m\}$$
$$= B_1 \cup B_2 \cup \ldots \cup B_m,$$

where the sets

$$B_j = \{r_1 + \ldots + r_{j-1} + 1, \ldots, r_1 + \ldots + r_{j-1} + r_j\}$$

are pairwise disjoint. (Convention: When $j = 1$, the 'empty sum' $r_1 + \ldots + r_{j-1}$ is understood to mean 0.)

For each $j = 1, \ldots, m$ the formula $\theta_j(i) = r_1 + \ldots + r_{j-1} + i$ defines a bijection

$$\theta_j : \{1, \ldots, r_j\} \to B_j.$$

Define $\sigma : \{1, \ldots, r\} \to A_1 \cup \ldots \cup A_m$ as follows. If $k \in \{1, \ldots, r\}$ then $k \in B_j$ for a unique $j \in \{1, \ldots, m\}$, so $k = \theta_j(i)$ for a unique $i \in \{1, \ldots, r_j\}$; define

$$\sigma(k) = \sigma_j(i) = \sigma_j(\theta_j^{-1}(k)).$$

In other words, σ is the unique mapping on $\{1, \ldots, r\}$ that agrees with $\sigma_j \circ \theta_j^{-1}$ on B_j.

It remains only to show that σ is surjective; indeed,

$$\sigma(\{1, \ldots, r\}) = \sigma(B_1 \cup \ldots \cup B_m)$$
$$= \sigma(B_1) \cup \ldots \cup \sigma(B_m)$$
$$= \sigma_1(\theta_1^{-1}(B_1)) \cup \ldots \cup \sigma_m(\theta_m^{-1}(B_m))$$
$$= \sigma_1(\{1, \ldots, r_1\}) \cup \ldots \cup \sigma_m(\{1, \ldots, r_m\})$$
$$= A_1 \cup \ldots \cup A_m. \quad ◊$$

4.4.6. **Lemma.** *If* $A \subset \mathbb{P}$ *and* A *is infinite, then there exists a strictly increasing mapping* $\varphi : \mathbb{P} \to A$; *in particular,* φ *is injective and* $\varphi(n) \geq n$ *for all* $n \in \mathbb{P}$.

Proof. Apart from notation, it's the same to show that there exists a sequence (a_n) in A such that $m < n \Rightarrow a_m < a_n$ (§2.5). Define a_n recursively as follows. Since A is not finite, it is nonempty; let a_1 be the smallest element of A.[3] Then $A \neq \{a_1\}$ (because $\{a_1\}$ is finite and A isn't), so $A - \{a_1\} \neq \varnothing$; let a_2 be the smallest element of $A - \{a_1\}$. Then $a_2 > a_1$ and $A \neq \{a_1, a_2\}$ (because $\{a_1, a_2\}$ is finite), so $A - \{a_1, a_2\}$ has a smallest element a_3, and $a_3 > a_2$. Assuming $a_1, \ldots, a_n$ already defined in this fashion, let a_{n+1} be the smallest element of $A - \{a_1, \ldots a_n\}$.

The function $\varphi : \mathbb{P} \to A$ defined by $\varphi(n) = a_n$ is strictly increasing. It follows, by induction, that $\varphi(n) \geq n$ for all n. For, $\varphi(1) \geq 1$; and if $\varphi(k) \geq k$ then $\varphi(k+1) > \varphi(k) \geq k$, therefore $\varphi(k+1) \geq k+1$. ◇

4.4.7. **Theorem.** *Every subset of a finite set is finite.*

Proof. Suppose F is finite and $B \subset F$. By assumption, there exists a surjection $\sigma : \{1, \ldots, r\} \to F$ for some positive integer r. Let $A = \sigma^{-1}(B)$ be the inverse image[4] of B under σ; then $\sigma(A) = B$ (because σ is surjective), so it will suffice to show that A is finite (4.4.4).

Now, $A \subset \{1, \ldots, r\}$. If A were infinite, by the lemma there would exist a mapping $\varphi : \mathbb{P} \to A$ such that $\varphi(n) \geq n$ for all $n \in \mathbb{P}$; but then $n \leq \varphi(n) \leq r$ for all n, which is absurd for $n = r+1$. ◇

4.4.8. **Corollary.** *Every superset of an infinite set is infinite.*

Proof. Suppose $B \supset A$. By the theorem,

$$B \text{ finite } \Rightarrow A \text{ finite};$$

in contrapositive form, this says

$$A \text{ not finite } \Rightarrow B \text{ not finite.} ◇$$

4.4.9. **Corollary.** *If* $\varphi : \mathbb{P} \to A$ *is injective, then* A *is infinite.*

Proof. If $B = \varphi(\mathbb{P})$ is the range of φ, then φ defines a bijection $\mathbb{P} \to B$; let $\psi : B \to \mathbb{P}$ be the inverse of this bijection. Since $\psi(B) = \mathbb{P}$ and $\mathbb{P}$ is infinite (4.4.3), B can't be finite (4.4.4); but $B \subset A$, so A isn't finite either (4.4.8). ◇

The property in 4.4.9 characterizes infinite sets (if you believe the following proof of the converse):

---

[3] We're citing here the 'well-ordering property' of the set of positive integers: every nonempty subset of $\mathbb{P}$ has a first element.

[4] The set of all $i \in \{1, \ldots, r\}$ such that $\sigma(i) \in B$ (see also §5.1).

4.4.10. Theorem. *A set* A *is infinite if and only if there exists an injection* $\mathbb{P} \to A$.

Proof. The "if" part is 4.4.9.

Conversely, assuming A infinite, we have to produce a sequence (a_n) in A such that $m \neq n \Rightarrow a_m \neq a_n$. "Construct" a_n recursively as follows.[5] Since A is infinite, it is nonempty; choose $a_1 \in A$. Then $A \neq \{a_1\}$ (because $\{a_1\}$ is finite), so $A - \{a_1\} \neq \varnothing$; choose $a_2 \in A - \{a_1\}$. Assuming $a_1, \ldots, a_n$ already chosen, $A \neq \{a_1, \ldots a_n\}$; choose $a_{n+1} \in A - \{a_1, \ldots, a_n\}$. ◊

Exercises

1. (i) Why is the set $\mathbb{Q}$ of rational numbers infinite?

(ii) Prove that the set $\mathbb{R} - \mathbb{Q}$ of irrational numbers is infinite. {Hint: Consider the mapping $n \mapsto n + \sqrt{2}$ $(n \in \mathbb{P})$.}

2. If $f : X \to Y$ is injective and X is infinite, prove that Y is infinite.

3. Prove that every interval with endpoints $a < b$ is an infinite subset of $\mathbb{R}$.

4. If A and B are sets then the *product set* $A \times B$ is the set of all ordered pairs (a, b) such that $a \in A$ and $b \in B$, with the understanding that $(a, b) = (a', b')$ means that $a = a'$ and $b = b'$.

(i) Prove that if A and B are finite then so is $A \times B$.

(ii) If A and B are nonempty and $A \times B$ is finite, prove that A and B are finite.

{Hints: (i) The correspondence $(i, j) \mapsto (i - 1)s + j$ defines a bijection

$$\{1, \ldots, r\} \times \{1, \ldots, s\} \to \{1, \ldots, rs\}.$$

(ii) Consider the 'projection mappings' $(a, b) \mapsto a$ and $(a, b) \mapsto b$.}

5. Let A be a subset of $\mathbb{R}$. A point $x \in \mathbb{R}$ is called a *limit point* of A if every neighborhood of x contains a point of A different from x. {For example, 0 is a limit point of each of the intervals $(0, 1)$ and $[0, 1)$.} Prove:

(i) If x is a limit point of A, then $x \in \overline{A}$.

(ii) If $x \in \overline{A} - A$ then x is a limit point of A.

(iii) x is a limit point of A $\Leftrightarrow$ $x \in \overline{A - \{x\}}$.

(iv) Every bounded infinite subset of $\mathbb{R}$ has at least one limit point. {Hint: Apply 4.4.10, then the Weierstrass–Bolzano theorem.}

---

[5]Superficially, the argument looks a lot like the proof of 4.4.6; the difference is that, without the well-ordering property used in the proof of 4.4.6, there is ambiguity in the choice of the a_n.

6. If A is a subset of $\mathbb{R}$, the set of all limit points of A (Exercise 5) is called the *derived set* of A, traditionally denoted A' (regrettably conflicting with a popular notation for 'complement'; we just have to keep our eye on the ball).

(i) If $x_n \to x$, $x_n \neq x$ ($\forall\, n$) and $A = \{x\} \cup \{x_n : n \in \mathbb{P}\}$, then $A' = \{x\}$.

(ii) If $A \subset B \subset \mathbb{R}$ then $A' \subset B'$.

(iii) $(A \cup B)' = A' \cup B'$ for all subsets A, B of $\mathbb{R}$.

(iv) If k is any positive integer, give an example of a set $A \subset \mathbb{R}$ such that A' has exactly k elements.

(v) Give an example with $A' = \{0\} \cup \{1/n : n \in \mathbb{P}\}$.

(vi) A' is a closed set; why?

(vii) $\overline{A} = A \cup A'$ for every subset A of $\mathbb{R}$.

4.5. Heine-Borel Covering Theorem

The Heine-Borel theorem is a theorem about open coverings of a closed interval; first, we have to explain the words.

4.5.1. *Definitions.* Let $A \subset \mathbb{R}$ and let $\mathcal{C}$ be a set of subsets of $\mathbb{R}$. If each point of A belongs to some set in $\mathcal{C}$, we say that $\mathcal{C}$ is a **covering of** A (or that $\mathcal{C}$ **covers** A); in symbols,

$$(\forall\, x \in A) \ \exists\, C \in \mathcal{C} \ni x \in C$$

or, more concisely,

$$A \subset \bigcup \mathcal{C}.$$

If, moreover, every set in $\mathcal{C}$ is an open subset of $\mathbb{R}$, then $\mathcal{C}$ is said to be an **open** covering of A. If a covering $\mathcal{C}$ of A consists of only a finite number of sets, it is called a **finite** covering. If $\mathcal{C}$ is a covering of A and if $\mathcal{D} \subset \mathcal{C}$ is such that $\mathcal{D}$ is also a covering of A, then $\mathcal{D}$ is referred to as a **subcovering** (it is a subset of $\mathcal{C}$ but a covering of A).

To get a feeling for these concepts, let's look at two instructive examples.

4.5.2. *Example.* Suppose A consists of the terms of a convergent sequence and its limit, that is,

$$A = \{x\} \cup \{x_n : n \in \mathbb{P}\},$$

where $x_n \to x$. If $\mathcal{C}$ is an open covering of A then A is covered by finitely many of the sets in $\mathcal{C}$. {Proof: The limit x belongs to one of the sets in $\mathcal{C}$, say $x \in U \in \mathcal{C}$. Since U is open, there is an $\epsilon > 0$ with $(x - \epsilon, x + \epsilon) \subset U$; it follows that $x_n \in U$ ultimately, say for $n > N$. Each of the terms x_i ($i = 1, \dots, N$) belongs to some $U_i \in \mathcal{C}$, so A is covered by the sets $U, U_1, \ldots, U_N$.} So to speak, every open covering of A admits a finite subcovering.

4.5.3. *Example.* Let A be the open interval $(2, 5)$ and let C be the set of all open intervals $(2 + \frac{1}{n}, 5 - \frac{1}{n})$ $(n \in \mathbb{P})$. Then C is an open covering of A, but no finite set of elements of C can cover A. {Reason: Each element of C is a proper subset of A and, among any finite set of elements of C, one of them contains all the others.} Thus, C is an open covering of A that admits no finite subcovering.

Without further fanfare:

4.5.4. **Theorem.** (Heine-Borel theorem) *If $[a, b]$ is a closed interval in $\mathbb{R}$ and C is an open covering of $[a, b]$, then $[a, b]$ is covered by a finite number of the sets in C.*

Proof. Let S be the set of all $x \in [a, b]$ such that the closed interval $[a, x]$ is covered by finitely many sets of C. At least $a \in S$, because $[a, a] = \{a\}$ and a belongs to some set in C; our objective is to show that $b \in S$.

At any rate, S is nonempty and bounded; let $m = \sup S$. Since $S \subset [a, b]$, we have $a \leq m \leq b$. The strategy of the proof is to show that (1) $m \in S$, and (2) $m = b$.

(1) Since $m \in [a, b] \subset \bigcup C$, there is a $V \in C$ such that $m \in V$; since V is open, $[m - \epsilon, m + \epsilon] \subset V$ for some $\epsilon > 0$. Thought (for later use): we can take ϵ to be as small as we like.

Since $m - \epsilon < m$ and m is the *least* upper bound of S, there exists $x \in S$ with $m - \epsilon < x \leq m$. From $x \in S$ we know that the interval $[a, x]$ is covered by finitely many sets in C, say

$$[a, x] \subset U_1 \cup \ldots \cup U_r;$$

on the other hand, $[x, m] \subset [m - \epsilon, m + \epsilon] \subset V$, so $[a, m] = [a, x] \cup [x, m]$ is covered by the sets $V, U_1, \ldots, U_r$ of C. This proves that $m \in S$, and a little more:

(∗) $[a, m + \epsilon] \subset V \cup U_1 \cup \ldots \cup U_r;$

it follows that $m + \epsilon > b$, because $m + \epsilon \leq b$ would imply, by virtue of (∗), that $m + \epsilon \in S$, contrary to the fact that every element of S is $\leq m$.

(2) The preceding argument shows that $b - m < \epsilon$ and the argument is valid with ϵ replaced by any positive number smaller than ϵ. It follows that $b - m \leq 0$. {The alternative, $0 < b - m < \epsilon$, would entail the absurdity $b - m < b - m$.} Thus $b \leq m$; already $m \leq b$, so $b = m \in S$. $\Diamond$

The Heine-Borel theorem states that if $A = [a, b]$ then every open covering of A admits a finite subcovering; the same is true for the set $A = \{x\} \cup \{x_n : n \in \mathbb{P}\}$ formed by a convergent sequence and its limit

(4.5.2). Two examples precipitate a definition (and a question—are there other examples?):

4.5.5. *Definition.* A subset A of $\mathbb{R}$ is said to be **compact** if every open covering of A admits a finite subcovering.

Such sets are readily characterized:

4.5.6. Theorem. *For a subset* A *of* $\mathbb{R}$, *the following conditions are equivalent*:

(a) A *is compact*;
(b) A *is bounded and closed*.

Proof. (a) $\Rightarrow$ (b): Suppose A is compact. The open intervals $(-n, n)$ ($n \in \mathbb{P}$) have union $\mathbb{R}$, so they certainly cover A; by hypothesis, a finite number of them suffice to cover A, which means that $A \subset (-m, m)$ for some m, consequently A is bounded.

To show that A is closed we need only show that $\overline{A} \subset A$ (§4.2), equivalently, $\complement A \subset \complement \overline{A}$. Assuming $x \notin A$ let's show that $x \notin \overline{A}$; we seek a neighborhood V of x such that $V \cap A = \varnothing$ (cf. 4.3.3). If $a \in A$ then $x \neq a$ (because $x \notin A$), so there exist open intervals U_a, V_a such that $a \in U_a$, $x \in V_a$ and $U_a \cap V_a = \varnothing$ (Figure 8).

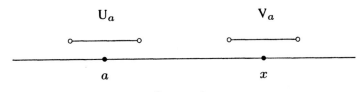

$$U_a \qquad\qquad\qquad\qquad V_a$$

$$a \qquad\qquad\qquad\qquad x$$

FIGURE 8

As a varies over A, the sets U_a form an open covering of A; suppose

$$A \subset U_{a_1} \cup \ldots \cup U_{a_r}.$$

Let

$$U = U_{a_1} \cup \ldots \cup U_{a_r}, \quad V = V_{a_1} \cap \ldots \cap V_{a_r}.$$

Then $A \subset U$ and V is a neighborhood of x (4.3.4). If $y \in U_{a_j}$ then $y \notin V_{a_j}$, therefore $y \notin V$; it follows that $V \cap U = \varnothing$ (V misses every term in the formula for U, so it misses their union), consequently $V \cap A = \varnothing$.

(b) $\Rightarrow$ (a): Assuming that A is bounded and closed and that $\mathcal{C}$ is an open covering of A, we seek a finite subcovering. By hypothesis, the set $V = \mathbb{R} - A$ is open and A is contained in some closed interval, say $A \subset [a, b]$. Why not apply the Heine-Borel theorem to $[a, b]$? The points of $[a, b]$ that are in A are covered by $\mathcal{C}$; what's left, $[a, b] - A$, is contained in V. We thus have an open covering of $[a, b]$: the sets in $\mathcal{C}$, helped out by the set V. It follows that

$(*)$ $$[a, b] \subset V \cup U_1 \cup \ldots \cup U_r$$

for suitable $U_1, \ldots, U_r$ in $\mathcal{C}$. The set A is contained in $[a, b]$ but is disjoint from V, so it follows from (*) that

$$A \subset U_1 \cup \ldots \cup U_r$$

and we have arrived at the desired finite subcovering. ◇

The concept of compactness is exceptionally fertile; we will have many occasions to appreciate it.

4.5.7. Corollary. *Every nonempty compact set* $A \subset \mathbb{R}$ *has a largest element and a smallest element.*

Proof. By the theorem, A is bounded and closed. Let $M = \sup A$ and choose a sequence (x_n) in A such that $x_n \to M$ (3.4.9); then $M \in A$ (because A is closed) and M is obviously the largest element of A. Similarly, $\inf A$ belongs to A and is its smallest element. ◇

Exercises

1. Let A be a subset of $\mathbb{R}$. Prove that the following conditions are equivalent:
  (a) A is compact;
  (b) every sequence in A has a subsequence that converges to a point of A.

2. Let A and B be closed subsets of $\mathbb{R}$, C a compact subset of $\mathbb{R}$. Prove:
  (i) The set $A + C = \{a + c : a \in A, c \in C\}$ is closed. {Hint: If $a_n + c_n \to x$, pass to a convergent subsequence of (c_n).}
  (ii) The set $A + B = \{a + b : a \in A, b \in B\}$ need not be closed. (Try $A = \{-1, -2, -3, \ldots\}$ and $B = \{n + 1/n : n = 2, 3, 4, \ldots\}$.)
  (iii) If $A \subset C$ then A is also compact.
  (iv) If A and B are compact, then so are $A + B$, $A \cap B$ and $A \cup B$.
  (v) If $\mathcal{K}$ is a set of closed subsets of $\mathbb{R}$, at least one of which is compact, then $\bigcap \mathcal{K}$ is compact.

3. True or false (explain): If A is any bounded subset of $\mathbb{R}$, then its derived set A' is compact. (Cf. §4.4, Exercise 6.)

4. True or false (explain): The only compact intervals in $\mathbb{R}$ are the closed intervals $[a, b]$.

5. Let A be a nonempty open set in $\mathbb{R}$. Prove:
  (i) A is the union of the set $\mathcal{C}$ of all open intervals (r, s) with r and s rational and $(r, s) \subset A$.
  (ii) A is the union of the set $\mathcal{D}$ of all open intervals (r, s) with r and s rational and $[r, s] \subset A$.

(iii) $\mathcal{C}$ and $\mathcal{D}$ are open coverings of A, $\mathcal{D} \subset \mathcal{C}$.

(iv) $\mathcal{D}$ admits no finite subcovering. {Hint: §4.3, Exercise 14.}

(v) $\mathcal{C}$ admits a finite subcovering if and only if A is the union of a finite number of open intervals with rational endpoints.

6. Let (X, d) be a metric space (§3.5, Exercise 8). Call a set $\mathcal{C}$ of subsets (open subsets) of X a *covering* (*open covering*) of X if $\bigcup \mathcal{C} = X$. It can be shown (the proof is not easy) that the following conditions are equivalent:

(a) Every sequence in X has a convergent subsequence.

(b) Every open covering of X admits a finite subcovering.

Thus, the definition of compactness mentioned in §3.5, Exercise 8 is equivalent to the definition suggested by 4.5.5.

7. If $x_n \to x$ and $A = \{x\} \cup \{x_n : n \in \mathbb{P}\}$, then A is compact (4.5.2) therefore it is closed and bounded by 4.5.6. Prove that A is closed and bounded without using 4.5.6.

8. If (A_n) is a sequence of nonempty, compact subsets of $\mathbb{R}$ such that $A_n \supset A_{n+1}$ for all n, prove that the intersection $A = \bigcap A_n$ is nonempty. {Hint: If $A = \varnothing$ then the sets $\complement A_n$ ($n \geq 2$) form an open covering of A_1.}

CHAPTER 5

Continuity

In elementary calculus, continuity and derivatives are painted with the same brush and the brush is called 'limits'; this is good for showing the core of unity in the processes of calculus. However, when we look at these concepts more closely, they are quite different: continuity has to do with the interaction between functions and open sets, whereas differentiability involves in addition the algebraic structure of the number field (from the very outset one considers 'difference quotients').

The present chapter is devoted to the basic notions of continuity of real-valued functions of a real variable (functions $f : S \to \mathbb{R}$, where S is a subset of $\mathbb{R}$); the next chapter emphasizes those aspects of continuity that depend on the order properties of $\mathbb{R}$. Since functions are in the forefront from now on, we begin with a discussion of the effect of a function $f : X \to Y$ on subsets of its initial set X and its final set Y.

5.1. Functions, Direct Images, Inverse Images

The fundamental definitions concerning functions are summarized in the Appendix (§A.3). A function $f : X \to Y$ acts on points of X to produce points of Y. In this chapter it will be useful to let f also act on subsets of X to produce subsets of Y, and *vice versa* (even if f does not have an inverse function in the sense of A.3.10!).

The basic idea is very simple: if A is a subset of X we can let f act on all of the elements of A; the result is a set of elements of Y, in other words a subset of Y, denoted $f(A)$ and called the *image* (or 'direct image') of A under f. In symbols,

$$f(A) = \{y \in Y : y = f(x) \text{ for some } x \in A\}$$
$$= \{f(x) : x \in A\}.$$

In particular, if A is a singleton, say $A = \{a\}$, then $f(A)$ is also a singleton:

$$f(\{a\}) = \{f(a)\}.$$

More generally, if $x_1, \ldots, x_n$ is any finite list of elements of X, then

$$f(\{x_1, \ldots, x_n\}) = \{f(x_1), \ldots, f(x_n)\}.$$

In the reverse direction (from Y to X) the situation (and the notation) is a little more complicated: if B is a subset of Y, we consider the elements x of X *that are mapped by f into* B, in other words, such that $f(x) \in B$. The set of all such elements x (there may not be any!) forms a subset of X (possibly empty), called the *inverse image of* B *under* f, denoted $f^{-1}(B)$; in symbols,

$$f^{-1}(B) = \{x \in X : f(x) \in B\}.$$

5.1.1. *Example.* Let $f : \mathbb{R} \to \mathbb{R}$ be the function $f(x) = x^2$. Then

$$f(\{2\}) = \{4\}, \quad f(\{-2, 2\}) = \{4\}, \quad f^{-1}(\{4\}) = \{-2, 2\}.$$

Also,

$$f([0, 2]) = [0, 4] = f([-1, 2]), \quad f^{-1}([0, 4]) = [-2, 2]$$

and $f([0, +\infty)) = [0, +\infty)$ (why?).

*5.1.2. *Example.* Let f be the sine function, that is, define $f : \mathbb{R} \to \mathbb{R}$ by $f(x) = \sin x$. Then $f(\pi) = 0$, $f^{-1}(\{0\}) = \{n\pi : n \in \mathbb{Z}\}$, and $f^{-1}(\{\pi\}) = \emptyset$ (why?). Also,

$$f([-\pi/2, \pi/2]) = [-1, 1] = f(\mathbb{R}),$$
$$f^{-1}([0, 1]) = \bigcup_{n \in \mathbb{Z}} [2n\pi, (2n+1)\pi].$$

{Exercise: Unravel all of these formulas; i.e., what do they say? Draw a picture!}*

The following theorem helps us get a grip on these concepts:

5.1.3. Theorem. *Let $f : X \to Y$ be any function.*
(1) *For subsets A_1, A_2 of X,*

$$A_1 \subset A_2 \implies f(A_1) \subset f(A_2).$$

($1'$) *For subsets* B_1, B_2 *of* Y,

$$B_1 \subset B_2 \;\Rightarrow\; f^{-1}(B_1) \subset f^{-1}(B_2).$$

(2) $f(A_1 \cup A_2) = f(A_1) \cup f(A_2)$ *for all subsets* A_1, A_2 *of* X.
($2'$) $f^{-1}(B_1 \cup B_2) = f^{-1}(B_1) \cup f^{-1}(B_2)$ *for all subsets* B_1, B_2 *of* Y.
(3) $f(f^{-1}(B)) \subset B$ *for every subset* B *of* Y.
($3'$) $f^{-1}(f(A)) \supset A$ *for every subset* A *of* X.
($4'$) $f^{-1}(B_1 \cap B_2) = f^{-1}(B_1) \cap f^{-1}(B_2)$ *for all subsets* B_1, B_2 *of* Y.
($5'$) $f^{-1}(Y - B) = X - f^{-1}(B)$ *for every subset* B *of* Y.

Proof. {There are no (4) and (5); the obvious formulas that come to mind are in general false.}

(1) Assuming $y \in f(A_1)$ we have to show that $y \in f(A_2)$. By assumption $y = f(x)$ for some $x \in A_1$; but $A_1 \subset A_2$ so x also belongs to A_2, thus $y = f(x) \in f(A_2)$.

($1'$) If $x \in f^{-1}(B_1)$ then $f(x) \in B_1 \subset B_2$, so $f(x) \in B_2$, in other words $x \in f^{-1}(B_2)$.

(2) For a point y in Y,

$$
\begin{aligned}
y \in f(A_1 \cup A_2) &\Leftrightarrow y = f(x) \text{ for some } x \text{ in } A_1 \cup A_2 \\
&\Leftrightarrow y = f(x) \text{ for some } x \text{ in } A_1 \text{ or in } A_2 \\
&\Leftrightarrow y \in f(A_1) \text{ or } y \in f(A_2) \\
&\Leftrightarrow y \in f(A_1) \cup f(A_2).
\end{aligned}
$$

($2'$) For a point x in X,

$$
\begin{aligned}
x \in f^{-1}(B_1 \cup B_2) &\Leftrightarrow f(x) \in B_1 \cup B_2 \\
&\Leftrightarrow f(x) \in B_1 \text{ or } f(x) \in B_2 \\
&\Leftrightarrow x \in f^{-1}(B_1) \text{ or } x \in f^{-1}(B_2) \\
&\Leftrightarrow x \in f^{-1}(B_1) \cup f^{-1}(B_2).
\end{aligned}
$$

(3) If $x \in f^{-1}(B)$ then $f(x) \in B$; thus $f(f^{-1}(B)) \subset B$.

($3'$) If $x \in A$ then $f(x) \in f(A)$, so $x \in f^{-1}(f(A))$; thus $A \subset f^{-1}(f(A))$.

($4'$) For a point x in X,

$$
\begin{aligned}
x \in f^{-1}(B_1 \cap B_2) &\Leftrightarrow f(x) \in B_1 \cap B_2 \\
&\Leftrightarrow f(x) \in B_1 \text{ and } f(x) \in B_2 \\
&\Leftrightarrow x \in f^{-1}(B_1) \text{ and } x \in f^{-1}(B_2) \\
&\Leftrightarrow x \in f^{-1}(B_1) \cap f^{-1}(B_2).
\end{aligned}
$$

(5') For a point x in X,

$$x \in f^{-1}(Y - B) \Leftrightarrow f(x) \in Y - B$$
$$\Leftrightarrow f(x) \notin B$$
$$\Leftrightarrow x \notin f^{-1}(B)$$
$$\Leftrightarrow x \in X - f^{-1}(B).$$

Another way of expressing this formula: $f^{-1}(\complement_Y B) = \complement_X f^{-1}(B)$, where, for example, $\complement_Y B = Y - B$ is the complement of B in Y (cf. A.2.7). $\Diamond$

Exercises

1. Let $f : X \to Y$ be a constant function, say $f(x) = c \in Y$ for all $x \in X$, let x_1 and x_2 be distinct points of X, and let $A_1 = \{x_1\}$, $A_2 = \{x_2\}$.
  (i) Compare $f(A_1 \cap A_2)$ and $f(A_1) \cap f(A_2)$.
  (ii) Compare $f(X - A_1)$ and $Y - f(A_1)$; also $f(X - A_1)$ and $f(X) - f(A_1)$.

2. Let $f : X \to Y$ be any function.
  (i) Prove that $f^{-1}(B_1 - B_2) = f^{-1}(B_1) - f^{-1}(B_2)$ for all subsets B_1, B_2 of Y.
  (ii) For a subset B of Y, prove:

$$f^{-1}(B) = \varnothing \Leftrightarrow B \cap f(X) = \varnothing.$$

3. Let $f : X \to Y$ be a function.
  (i) To say that f is injective means that for every $y \in Y$, $f^{-1}(\{y\})$ is either empty or a singleton.
  (ii) To say that f is surjective means that for every $y \in Y$, $f^{-1}(\{y\})$ is nonempty.
  (iii) To say that f is bijective means ... (complete the sentence).

4. Let $f : X \to Y$ be an *injective* function. Prove:
  (i) $f^{-1}(f(A)) = A$ for every subset A of X.
  (ii) If A_1, A_2 are subsets of X such that $f(A_1) = f(A_2)$, then $A_1 = A_2$.
  (iii) $f(A_1 \cap A_2) = f(A_1) \cap f(A_2)$ for all subsets A_1, A_2 of X; more generally,
  (iv) if $\mathcal{S}$ is any set of subsets of X, then

$$f\left(\bigcap \mathcal{S}\right) = \bigcap f(\mathcal{S}),$$

where $\bigcap \mathcal{S} = \{x \in X : x \in A \text{ for all } A \in \mathcal{S}\}$ and $f(\mathcal{S}) = \{f(A) : A \in \mathcal{S}\}$.

(v) $f(A_1 - A_2) = f(A_1) - f(A_2)$ for all subsets A_1, A_2 of X.

5. If $f : X \to Y$ is a *surjective* function, prove that $B = f(f^{-1}(B))$ for every subset B of Y.

6. Every function $f : X \to Y$ can be written as a composite $f = i \circ s$ with s surjective and i injective. {Hint: Let $i : f(X) \to Y$ be the mapping $i(y) = y$, which 'inserts' the subset $f(X)$ of Y into Y.}

7. Let X and Y be sets, $\mathcal{P}(X)$ and $\mathcal{P}(Y)$ their power sets (A.2.8). A function $f : X \to Y$ defines functions $f_* : \mathcal{P}(X) \to \mathcal{P}(Y)$ and $f^* : \mathcal{P}(Y) \to \mathcal{P}(X)$, by the formulas

$$f_*(A) = f(A) = \{ f(x) : x \in A \},$$
$$f^*(B) = f^{-1}(B) = \{ x \in X : f(x) \in B \}.$$

Discuss the composite functions $f_* \circ f^*$ and $f^* \circ f_*$. {Hint: (3) and (3') of 5.1.3; cf. Exercises 4 and 5.}

5.2. Continuity at a Point

Continuity has to do with functions and open sets. The open sets are the complements of the closed sets, and our definition of closed set is based on the notion of convergent sequence (4.2.1), so continuity has to do with functions and convergent sequences; the precise definition is as follows:

5.2.1. *Definition.* Let $f : S \to \mathbb{R}$, where S is a subset of $\mathbb{R}$, and let $a \in S$. {In other words, a is a point of the domain of a real-valued function of a real variable.} We say that f is **continuous at** a if it has the following property:

$$x_n \in S, \ x_n \to a \ \Rightarrow \ f(x_n) \to f(a).$$

That is, if (x_n) is any sequence in S converging to the point a of S, then $(f(x_n))$ converges to $f(a)$. If f is *not* continuous at a, it is said to be **discontinuous at** a.

5.2.2. *Example.* The **identity function** $\mathrm{id}_{\mathbb{R}} : \mathbb{R} \to \mathbb{R}$ (cf. A.3.2) is continuous at every $a \in \mathbb{R}$.

5.2.3. *Example.* If $f : \mathbb{R} \to \mathbb{R}$ is a **constant function**, say $f(x) = c \ (\forall \, x \in \mathbb{R})$, then f is continuous at every $a \in \mathbb{R}$.

5.2.4. *Example.* The function $f : \mathbb{R} \to \mathbb{R}$ defined by

$$f(x) = \begin{cases} 1 & \text{for } x > 0 \\ 0 & \text{for } x \leq 0 \end{cases}$$

is discontinuous at $a = 0$. {Consider the sequence $x_n = 1/n$.}

5.2.5. *Example*. The function $f : \mathbb{R} \to \mathbb{R}$ defined by

$$f(x) = \begin{cases} 1 & \text{if } x \in \mathbb{Q} \\ 0 & \text{if } x \notin \mathbb{Q} \end{cases}$$

(where $\mathbb{Q}$ is the set of rational numbers) is discontinuous at every $a \in \mathbb{R}$. {If a is rational, consider the sequence $x_n = a + (1/n)\sqrt{2}$; if a is irrational, let x_n be a rational number with $a < x_n < a + 1/n$ (2.4.1).}

The definition of continuity (5.2.1) says it with sequences; we can also say it with epsilons:

5.2.6. **Theorem.** *Let $a \in S \subset \mathbb{R}$, $f : S \to \mathbb{R}$. The following conditions on f are equivalent:*
(a) *f is continuous at a;*
(b) *for every $\epsilon > 0$, there exists a $\delta > 0$ such that*

$$x \in S, \ |x - a| < \delta \ \Rightarrow \ |f(x) - f(a)| < \epsilon.$$

Proof. Just for practice, here is condition (b) in 'formal symbolese':

$$(\forall \, \epsilon > 0) \, \exists \, \delta > 0 \ \ni \ (x \in S \ \& \ |x - a| < \delta) \ \Rightarrow \ |f(x) - f(a)| < \epsilon.$$

What it says, informally, is that $f(x)$ is near $f(a)$ provided $x \in S$ is sufficiently near a. The degree of nearness to $f(a)$ (namely ϵ) is specified in advance (and arbitrary); the degree of nearness to a (namely δ) has to be found. If a smaller ϵ is specified, the chances are that δ will also have to be taken smaller (but not necessarily; for a constant function, whatever the given $\epsilon > 0$, *any* $\delta > 0$ will do). Now for the proof:
(b) $\Rightarrow$ (a): Let $x_n \in S$, $x_n \to a$; we have to show that $f(x_n) \to f(a)$. Let $\epsilon > 0$; we seek an index N such that $n \geq N \Rightarrow |f(x_n) - f(a)| < \epsilon$. Choose $\delta > 0$ as in (b) (to go along with the given ϵ), then choose N so that $n \geq N \Rightarrow |x_n - a| < \delta$ (possible because $x_n \to a$); in view of the implication in (b), $n \geq N \Rightarrow |f(x_n) - f(a)| < \epsilon$.
$\sim$(b) $\Rightarrow$ $\sim$(a): (We are proving (a) $\Rightarrow$ (b) in contrapositive form.) Assuming $\sim$(b), *there exists an $\epsilon > 0$ such that for every $\delta > 0$* the implication in (b) fails; thus

$$(*) \qquad (\forall \, \delta > 0) \, \exists \, x \in S \ \ni \ |x - a| < \delta \ \& \ |f(x) - f(a)| \geq \epsilon.$$

{What's going on? Condition (b) says that for every $\epsilon > 0$ there exists a successful $\delta > 0$, where success means that the implication in (b) is true. The *negation* of (b) says that there exists an $\epsilon > 0$ for which every $\delta > 0$ fails; for δ to fail means that the implication in (b) fails, and this means there is an x for which the left side of the implication is true but the right side is not.}

For each $n \in \mathbb{P}$ choose $x_n \in S$ so that $|x_n - a| < 1/n$ and $|f(x_n) - f(a)| \geq \epsilon$ (apply (*), with $\delta = 1/n$). Then (x_n) is a sequence in S that converges to a, but $(f(x_n))$ does not converge to $f(a)$; the existence of such a sequence negates the condition defining continuity at a (5.2.1). Assuming $\sim$(b), we have verified $\sim$(a). $\Diamond$

When the domain of f is all of $\mathbb{R}$, we can say it with neighborhoods:

5.2.7. Theorem. *If $f : \mathbb{R} \to \mathbb{R}$ and $a \in \mathbb{R}$, the following conditions are equivalent:*
(a) *f is continuous at a;*
(b) *for every neighborhood V of $f(a)$, $f^{-1}(V)$ is a neighborhood of a.*

Proof. (a) $\Rightarrow$ (b): Let V be a neighborhood of $f(a)$; according to 4.3.1, there is an $\epsilon > 0$ such that $(f(a) - \epsilon, f(a) + \epsilon) \subset V$. Since f is continuous, we know from the preceding theorem that there exists a $\delta > 0$ such that $|x - a| < \delta \Rightarrow |f(x) - f(a)| < \epsilon$, in other words

$$x \in (a - \delta, a + \delta) \Rightarrow f(x) \in (f(a) - \epsilon, f(a) + \epsilon).$$

Thus,
$$f\big((a - \delta, a + \delta)\big) \subset (f(a) - \epsilon, f(a) + \epsilon) \subset V,$$

whence $(a - \delta, a + \delta) \subset f^{-1}(V)$, which shows that $f^{-1}(V)$ is a neighborhood of a.

(b) $\Rightarrow$ (a): Let us verify criterion (b) of 5.2.6: given any $\epsilon > 0$, we seek a suitable $\delta > 0$. Since $V = (f(a) - \epsilon, f(a) + \epsilon)$ is a neighborhood of $f(a)$, by hypothesis $f^{-1}(V)$ is a neighborhood of a, so there exists a $\delta > 0$ such that $(a - \delta, a + \delta) \subset f^{-1}(V)$; this inclusion means that $x \in (a - \delta, a + \delta) \Rightarrow f(x) \in V$, in other words, $|x - a| < \delta \Rightarrow |f(x) - f(a)| < \epsilon$. $\Diamond$

Exercises

1. Show that the function $f : \mathbb{R} \to \mathbb{R}$ defined by $f(x) = |x|$ is continuous at every $a \in \mathbb{R}$. {Hint: 3.4.8, (5).}

2. The constant function $g : \mathbb{Q} \to \mathbb{R}$ defined by $g(r) = 1$ ($\forall r \in \mathbb{Q}$) is continuous at every $a \in \mathbb{Q}$; does this conflict with 5.2.5?

3. Let $f : \mathbb{R} \to \mathbb{R}$ be the bracket function $f(x) = [x]$ (see 2.3.2). Discuss the points of continuity and discontinuity of f. {To show that a function is discontinuous at a, exhibit an offending sequence $x_n \to a$.}

4. Let $f : S \to \mathbb{R}$, $a \in S \subset \mathbb{R}$. Suppose f has the property

$$x_n \in S, \ x_n \to a \Rightarrow (f(x_n)) \text{ is convergent.}$$

Prove that f is continuous at a.

{Hint: Interlace a sequence (x_n) converging to a with the constant sequence $a, a, a, \ldots$.}

5. Let A be a nonempty, proper subset of $\mathbb{R}$ (that is, $A \subset \mathbb{R}$, $A \neq \varnothing$, $A \neq \mathbb{R}$). Define $f : \mathbb{R} \to \mathbb{R}$ by

$$f(x) = \begin{cases} 1 & \text{if } x \in A \\ 0 & \text{if } x \notin A. \end{cases}$$

Prove that f has at least one point of discontinuity.

{Hint: If some point $a \in A$ is the limit of points of CA, then f is discontinuous at a. If no such point exists, then A is open (4.3.3); but then A can't be closed (§4.3, Exercise 14), so some point $b \in CA$ is a limit of points of A.}

6. Define $f : [0,1] \to [0,1]$ as follows. Let $x \in [0,1]$. If x is irrational, define $f(x) = 0$. If x is rational and $x \neq 0$, write $x = m/n$ with m and n positive integers having no common factor, and define $f(x) = 1/n$. Finally, define $f(0) = 0$.

Prove that f is continuous at every irrational point and at 0, discontinuous at every nonzero rational.

{Hint: If x is irrational and $r_k = m_k/n_k$ is a sequence of 'reduced' fractions with $r_k \to x$, imagine the consequences of (n_k) being a bounded sequence.}

7. Let $f : \mathbb{R} \to \mathbb{R}$ be the function defined by

$$f(x) = \begin{cases} x & \text{if } x \text{ is irrational} \\ 0 & \text{if } x \text{ is rational.} \end{cases}$$

Prove that f is continuous at 0, and discontinuous at every other point of $\mathbb{R}$.

8. Let (X, d) and (Y, D) be metric spaces, $f : X \to Y$, $a \in X$. The function f is said to be *continuous at a* if

$$x_n \in X, \ x_n \to a \ \Rightarrow \ f(x_n) \to f(a),$$

in other words (§3.4, Exercise 12),

$$d(x_n, a) \to 0 \ \Rightarrow \ D(f(x_n), f(a)) \to 0.$$

With neighborhoods defined as in §4.3, Exercise 10, prove that the following conditions are equivalent:

(a) f is continuous at a;

(b) $(\forall\, \epsilon > 0)\ \exists\, \delta > 0\ \ni$

$$d(x, a) < \delta\ \Rightarrow\ D(f(x), f(a)) < \epsilon\,;$$

(c) V a neighborhood of $f(a)$ in Y $\Rightarrow$ $f^{-1}(V)$ is a neighborhood of a in X.

5.3. Algebra of Continuity

Various algebraic combinations of continuous functions are continuous:

5.3.1. Theorem. *Suppose $a \in S \subset \mathbb{R}$ and $f : S \to \mathbb{R}$, $g : S \to \mathbb{R}$; let c be any real number.*

If f and g are continuous at a, then so are the functions $f + g$, fg and cf.

Proof. The functions in question are defined on S by the formulas

$$(f + g)(x) = f(x) + g(x)$$
$$(fg)(x) = f(x)g(x)$$
$$(cf)(x) = cf(x)$$

(that is, they are the 'pointwise' sum, product and scalar multiple).

If $x_n \in S$ and $x_n \to a$, then $f(x_n) \to f(a)$ and $g(x_n) \to g(a)$ by the assumptions on f and g, therefore

$$(f + g)(x_n) = f(x_n) + g(x_n) \to f(a) + g(a) = (f + g)(a)$$

by 3.4.8, (3). This shows that $f + g$ is continuous at a, and the proofs for fg and cf are similar. {Incidentally, cf is the special case of fg when g is the constant function equal to c.} ◊

This simple theorem pays immediate dividends:

5.3.2. Corollary. *Every polynomial function $p : \mathbb{R} \to \mathbb{R}$ is continuous at every point of $\mathbb{R}$.*

Proof. Say $p(x) = a_0 + a_1 x + a_2 x^2 + \ldots + a_r x^r$ $(x \in \mathbb{R})$, where the coefficients $a_0, a_1, \ldots, a_r$ are fixed real numbers. If $u : \mathbb{R} \to \mathbb{R}$ is the identity function $u(x) = x$, then p is a linear combination of powers of u:

$$p = a_0 1 + a_1 u + a_2 u^2 + \ldots + a_r u^r$$

(for example, u^3 is the function $x \mapsto (u(x))^3 = x^3$; by convention, u^0 is the constant function 1). Since u is continuous, so are its powers (5.3.1), therefore so is any linear combination of them (5.3.1 again). ◊

Quotients are a little more delicate (we have to avoid zero denominators):

5.3.3. Theorem. *With notations as in 5.3.1, assume that f and g are continuous at a and that g is not zero at any point of S. Then f/g is also continuous at a.*

Proof. The formula defining $f/g : S \to \mathbb{R}$ is

$$(f/g)(x) = f(x)/g(x) \quad (x \in S).$$

(In general, f/g is defined—by the same formula—on the subset $T = \{x \in S : g(x) \neq 0\}$ of S; in the present case, $T = S$.) If $x_n \in S$ and $x_n \to a$, then

$$(f/g)(x_n) = f(x_n)/g(x_n) \to f(a)/g(a) = (f/g)(a)$$

by 3.4.8, (7). $\Diamond$

5.3.4. *Example.* Suppose $p : \mathbb{R} \to \mathbb{R}$ and $q : \mathbb{R} \to \mathbb{R}$ are polynomial functions, q not the zero polynomial; let

$$F = \{x \in \mathbb{R} : q(x) = 0\},$$

which is a finite set (possibly empty). {By the *factor theorem* of elementary algebra, $q(c) = 0$ if and only if the linear polynomial $x - c$ is a factor of q, that is, $q(x) = (x - c)q_1(x)$ for a suitable polynomial q_1 and for all $x \in \mathbb{R}$. Thus every root of q splits off a linear factor, so the degree of q puts an upper bound on the number of roots.} Let $r = p/q$ be the quotient function (called a *rational function*), defined on the set $S = \mathbb{R} - F$ by the formula

$$r(x) = p(x)/q(x) \quad (x \in S).$$

If $f = p|S$ and $g = q|S$ are the restrictions of p and q to S (A.3.4), it is clear from 5.3.2 that f and g are continuous at every point of S, therefore so is $r = f/g$ (5.3.3).

Exercises

1. With notations as in 5.3.1, assume f and g are continuous at a. Define $|f| : S \to \mathbb{R}$ by the formula $|f|(x) = |f(x)|$ $(x \in S)$, and $h : S \to \mathbb{R}$ by the formula $h(x) = \max\{f(x), g(x)\}$. One writes $h = \sup(f, g)$, or $h = \max(f, g)$, even though h is not "the larger of f and g". Similarly $k = \inf(f, g) = \min(f, g)$ is defined by the formula $k(x) = \min\{f(x), g(x)\}$. Prove:
   (i) $|f|$ is continuous at a.
   (ii) h and k are continuous at a.

{Hint: (i) 3.4.8, (5). (ii) $h = \frac{1}{2}(f + g + |f - g|)$; cf. §2.9, Exercise 1.}

2. Notations as in 5.3.1. True or false (explain):
(i) If $f + g$ is continuous at a, then f and g are continuous at a.
(ii) If $|f|$ is continuous at a, then f is continuous at a.

3. Notations as in 5.3.3. True or false (explain):
(i) If f/g and g are continuous at a, then f is continuous at a.
(ii) If f/g and f are continuous at a, then g is continuous at a.

4. Find a rational function $r : \mathbb{R} \to \mathbb{R}$ that is not a polynomial function. (The focus of the problem is on the domain of r.)

5.4. Continuous Functions

5.4.1. Definition. Suppose $f : S \to \mathbb{R}$, where S is a subset of $\mathbb{R}$; f is said to be a **continuous function** (or 'continuous mapping') if it is continuous at *every* $a \in S$.

The polynomial and rational functions discussed in the preceding section are important examples of continuous functions (5.3.2, 5.3.4). An example not covered by these is the function $x \mapsto |x|$ (see also the exercises); here is another:

5.4.2. Example. The function $f : [0, \infty) \to \mathbb{R}$ defined by $f(x) = \sqrt{x}$ is continuous. {Proof: For a sequence (x_n) in $[0, \infty)$, it is clear that (x_n) is null if and only if $(\sqrt{x_n})$ is null (cf. §3.3, Exercise 2); this assures continuity at 0. If $x > 0$ and $x_n \to x$, substitute x_n for y in the inequality

$$|y - x| = |(\sqrt{y} - \sqrt{x})(\sqrt{y} + \sqrt{x})| \geq |\sqrt{y} - \sqrt{x}|\sqrt{x},$$

valid for all $x > 0$, $y > 0$.}

5.4.3. Theorem. *Let S be a nonempty subset of $\mathbb{R}$. If $f : S \to \mathbb{R}$, $g : S \to \mathbb{R}$ are continuous functions and c is any real number, then the functions $f + g$, fg and cf are also continuous; if, moreover, g is not zero at any point of S, then f/g is also continuous.*

Proof. This is immediate from 5.3.1 and 5.3.3. ◇

When the domain of f is all of $\mathbb{R}$, there is a neat formulation of continuity in terms of open sets (or closed sets):

5.4.4. Theorem. *For a function $f : \mathbb{R} \to \mathbb{R}$, the following conditions are equivalent:*
(a) *f is continuous;*
(b) *U open $\Rightarrow$ $f^{-1}(U)$ open;*
(c) *A closed $\Rightarrow$ $f^{-1}(A)$ closed.*

Proof. (a) $\Rightarrow$ (c): Assuming f is continuous at every point of $\mathbb{R}$, we have to show that the inverse image of any closed set is closed. Let A be a closed subset of $\mathbb{R}$; assuming $x_n \in f^{-1}(A)$ and $x_n \to x \in \mathbb{R}$, we have to show that $x \in f^{-1}(A)$. Since f is continuous, $f(x_n) \to f(x)$; but $f(x_n) \in A$ and A is closed, so $f(x) \in A$, in other words $x \in f^{-1}(A)$.

(c) $\Rightarrow$ (b): If U is an open set, its complement $\complement U$ is closed (4.3.5), therefore $f^{-1}(\complement U)$ is closed by (c); then $f^{-1}(\complement U) = \complement f^{-1}(U)$ shows that $f^{-1}(U)$ is the complement of a closed set, so $f^{-1}(U)$ is open.

(b) $\Rightarrow$ (a): Given any $a \in \mathbb{R}$, we have to show that f is continuous at a. Let $\epsilon > 0$; we seek a $\delta > 0$ such that $|x-a| < \delta \Rightarrow |f(x)-f(a)| < \epsilon$, that is,

$$x \in (a - \delta, a + \delta) \Rightarrow f(x) \in (f(a) - \epsilon, f(a) + \epsilon),$$

in other words $(a - \delta, a + \delta) \subset f^{-1}\big((f(a) - \epsilon, f(a) + \epsilon)\big)$. The interval $U = (f(a) - \epsilon, f(a) + \epsilon)$ is an open set, so $f^{-1}(U)$ is open by (b); obviously $f(a) \in U$, so $a \in f^{-1}(U)$, and the existence of a $\delta > 0$ such that $(a - \delta, a + \delta) \subset f^{-1}(U)$ follows from the fact that $f^{-1}(U)$ is a neighborhood of a. $\Diamond$

Exercises

1. If $f : \mathbb{R} \to \mathbb{R}$ is continuous and U is an open set in $\mathbb{R}$, then $f(U)$ need not be an open set (consider a constant function). Find an example of a continuous function $f : \mathbb{R} \to \mathbb{R}$ and a closed set A such that $f(A)$ is not closed. {Hint: Find the range of the function $f(x) = x^2/(1+x^2)$.}

2. If $f : [0,1] \to [0,1]$ is continuous, then f has a fixed point, that is, there exists a point $a \in [0,1]$ such that $f(a) = a$.
{Hint: Check that the set $A = \{x \in [0,1] : f(x) \geq x\}$ is nonempty and let $a = \sup A$.}

3. If $f : [a,b] \to \mathbb{R}$ is continuous, prove that there exists a continuous function $F : \mathbb{R} \to \mathbb{R}$ that 'extends' f, in the sense that $F(x) = f(x)$ for all $x \in [a,b]$. {Hint: For $x > b$ define $F(x) = f(b)$.}

4. Let $f : \mathbb{R} \to \mathbb{R}$. The *graph* of f is the set

$$G_f = \{(x, f(x)) : x \in \mathbb{R}\} = \{(x,y) : x \in \mathbb{R} \ \& \ y = f(x)\}.$$

(i) Suppose f is continuous. Prove that if $(x_n, y_n) \in G_f$ and $x_n \to x$, $y_n \to y$, then $y = f(x)$, that is, $(x,y) \in G_f$. Does this remind you of closed sets?

(ii) Find a discontinuous function whose graph has the property in (i). {Hint: $1/x$.}

5. If $f : \mathbb{R} \to \mathbb{R}$ is continuous and A is a compact subset of $\mathbb{R}$, then $f(A)$ is also compact. {Hint: If C is an open covering of $f(A)$, then the sets $f^{-1}(U)$, where $U \in C$, form an open covering of A.}

6. Let A be a nonempty subset of $\mathbb{R}$ and define $f : \mathbb{R} \to \mathbb{R}$ by the formula
$$f(x) = \inf\{|x - a| : a \in A\}.$$
Prove that f is a continuous function. {Hint: $|x - a| \le |x - y| + |y - a|$; show that $|f(x) - f(y)| \le |x - y|$.}

7. If $f : [a, b] \to \mathbb{R}$ is continuous, then the range of f is a compact subset of $\mathbb{R}$. {Hint: Exercises 3 and 5.}

8. Prove: (i) There exists no continuous surjection $f : [-1, 1] \to (-1, 1)$. {Hint: Exercise 7.}

(ii) There exists a continuous bijection $g : \mathbb{R} \to (-1, 1)$ with continuous inverse. {Hint: Consider $g(x) = x/(1 + |x|)$.}

(iii) There exists a continuous surjection $h : \mathbb{R} \to [-1, 1]$.

(iv) There exists a continuous surjection $k : (-1, 1) \to [-1, 1]$.

9. If I is an interval in $\mathbb{R}$ and $f : I \to \mathbb{R}$ is continuous, then $f(I)$ is also an interval (*Intermediate value theorem*).

{Hint: If $I = \mathbb{R}$, argue that $f(\mathbb{R})$ is connected (§4.3, Exercise 17). If $I = [a, b]$ then f is extendible to a continuous function $F : \mathbb{R} \to \mathbb{R}$ with $F(\mathbb{R}) = f(I)$ (Exercise 3). In general, I is the union of an increasing sequence of closed intervals; cf. §4.3, Exercise 17, (iii).}

10. If $f : \mathbb{R} \to \mathbb{R}$ and $g : \mathbb{R} \to \mathbb{R}$ are continuous and $f(r) = g(r)$ for every rational number r, then $f = g$. {The role of the set $\mathbb{Q}$ of rational numbers can be played by any dense subset of $\mathbb{R}$ (cf. §4.3, Exercise 7).}

11. Let A be a closed subset of $\mathbb{R}$ and suppose $c \notin A$. Construct a continuous function $f : \mathbb{R} \to [0, 1]$ such that $f(c) = 0$ and $f(x) = 1$ for all $x \in A$ (one can even arrange for f to be zero in a neighborhood of c). {Hint: $\mathbb{R} - A$ contains an open interval containing c.}

5.5. One-Sided Continuity

In discussing functions f defined on an interval $[a, b]$, behavior at the endpoints requires some special treatment (for example, the point a can only be approached from the right); that's one reason for considering 'one-sided' matters. Another reason is that 'two-sided' behavior can often profitably be discussed by breaking it up into 'left-behavior' and 'right-behavior'. For example, the function f with the graph of Figure 9 has a tangent line problem at the origin; as x approaches 0 from the right, the slope of the chord joining $(0, 0)$ and $(x, f(x))$ approaches 1, and for x

approaching 0 from the left, the slope of the chord approaches -1; but the function fails to have a well-defined 'slope' at $(0,0)$ because the 'left slope' and 'right slope' are different.

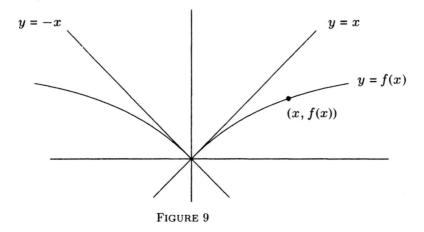

$y = -x$ $y = x$

$y = f(x)$

$(x, f(x))$

FIGURE 9

In this section we lay the technical foundation for one-sided discussions of this sort.

5.5.1. *Definition.* Let $a \in \mathrm{N} \subset \mathbb{R}$. We say that N is a **right neighborhood** of a if there exists an $r > 0$ such that $[a, a+r] \subset \mathrm{N}$. If there exists an $s > 0$ such that $[a-s, a] \subset \mathrm{N}$ then N is called a **left neighborhood** of a. {Thus N is a neighborhood of a in the sense of 4.3.1 if and only if it is both a left neighborhood and a right neighborhood of a.}

For example, if $a < b$ then the closed interval $[a, b]$ is a right neighborhood of a, a left neighborhood of b, and a neighborhood of each of its internal points $x \in (a, b)$.

5.5.2. If M and N are right neighborhoods of a, then so is $\mathrm{M} \cap \mathrm{N}$. If M is a right neighborhood of a and $\mathrm{M} \subset \mathrm{N}$ then N is also a right neighborhood of a.

5.5.3. *Definition.* Let $a \in \mathrm{S} \subset \mathbb{R}$, $f : \mathrm{S} \to \mathbb{R}$. We say that f is **right continuous** at a if (i) S is a right neighborhood of a, and (ii) if (x_n) is a sequence in S such that $x_n > a$ and $x_n \to a$, then $f(x_n) \to f(a)$, in symbols,

$$\left.\begin{array}{l} x_n \in \mathrm{S} \\ x_n > a \\ x_n \to a \end{array}\right\} \;\Rightarrow\; f(x_n) \to f(a).$$

'Left continuity' is defined dually (with "right" replaced by "left" and "$x_n > a$" by "$x_n < a$").

5.5.4. Suppose $a \in S \subset \mathbb{R}$, $f : S \to \mathbb{R}$. Let $T = -S = \{-x : x \in S\}$ and define $g : T \to \mathbb{R}$ by $g(x) = f(-x)$. Then f is left continuous at a if and only if g is right continuous at $-a$.

5.5.5. Theorem. *Suppose $f : S \to \mathbb{R}$ and S is a right neighborhood of a. The following conditions on f are equivalent:*
(a) f *is right continuous at a;*
(b) *for every $\epsilon > 0$ there exists a $\delta > 0$ such that*

$$a < x < a + \delta \implies |f(x) - f(a)| < \epsilon;$$

(c) V *a neighborhood of $f(a)$ $\implies$ $f^{-1}(V)$ is a right neighborhood of a.*

Proof. (a) $\implies$ (b): Let's prove $\sim$(b) $\implies$ $\sim$(a). The argument is similar to the one given in 5.2.6: condition (b) says that for every $\epsilon > 0$ there is a 'successful' $\delta > 0$; its negation means that there exists an $\epsilon_0 > 0$ for which every $\delta > 0$ 'fails'. In particular, for each $n \in \mathbb{P}$, $\delta = 1/n$ fails, so there exists a point $x_n \in S$ with $a < x_n < a + 1/n$, such that $|f(x_n) - f(a)| \geq \epsilon_0$. Then $x_n > a$ and $x_n \to a$ but $(f(x_n))$ does not converge to $f(a)$, so f is not right continuous at a.
(b) $\implies$ (c): If V is a neighborhood of $f(a)$, there is an $\epsilon > 0$ such that $(f(a) - \epsilon, f(a) + \epsilon) \subset V$. Choose $\delta > 0$ as in (b). By the implication in (b),

$$f((a, a + \delta)) \subset (f(a) - \epsilon, f(a) + \epsilon) \subset V.$$

Also $f(a) \in V$, so $f([a, a+\delta)) \subset V$; thus $[a, a+\delta) \subset f^{-1}(V)$, so $f^{-1}(V)$ is a right neighborhood of a.
(c) $\implies$ (a): Assuming $x_n \in S$, $x_n > a$, $x_n \to a$, we have to show that $f(x_n) \to f(a)$. Let $\epsilon > 0$; we must show that $|f(x_n) - f(a)| < \epsilon$ ultimately. Since $V = (f(a) - \epsilon, f(a) + \epsilon)$ is a neighborhood of $f(a)$, by hypothesis $f^{-1}(V)$ is a right neighborhood of a, so there is a $\delta > 0$ such that $[a, a + \delta) \subset f^{-1}(V)$; ultimately $x_n \in [a, a + \delta)$, so $f(x_n) \in V$, that is, $|f(x_n) - f(a)| < \epsilon$. $\Diamond$

The relation between continuity and 'one-sided continuity' is as follows:

5.5.6. Theorem. *If $a \in S \subset \mathbb{R}$ and $f : S \to \mathbb{R}$, the following conditions are equivalent:*
(a) f *is both left and right continuous at a;*
(b) S *is a neighborhood of a and f is continuous at a.*

Proof. (a) $\implies$ (b): By the definition of 'one-sided continuity' (5.5.3), S is both a left and right neighborhood of a, hence is a neighborhood of a. If V is a neighborhood of $f(a)$, then $f^{-1}(V)$ is both a left neighborhood and a right neighborhood of a (5.5.5 and its dual), hence is a neighborhood of a. In particular, if $\epsilon > 0$ and $V = (f(a) - \epsilon, f(a) + \epsilon)$, then there exists a $\delta > 0$ such that $(a - \delta, a + \delta) \subset f^{-1}(V)$, that is, $|x - a| < \delta \implies |f(x) - f(a)| < \epsilon$. Thus f is continuous at a (5.2.6).

(b) $\Rightarrow$ (a): By assumption, S is a neighborhood of a, and

$$x_n \in \mathrm{S}, \ x_n \to a \ \Rightarrow \ f(x_n) \to f(a).$$

In particular, S is a right neighborhood of a and

$$x_n \in \mathrm{S}, \ x_n > a, \ x_n \to a \ \Rightarrow \ f(x_n) \to f(a),$$

thus f is right continuous at a. Similarly, f is left continuous at a. $\Diamond$

5.5.7. Corollary. *If* $f : [a, b] \to \mathbb{R}$, $a < b$, *then the following conditions are equivalent*:
(a) f *is continuous*;
(b) f *is right continuous at* a, *left continuous at* b, *and both left and right continuous at each point of the open interval* (a, b).

Proof. The first statement in (b) means that f is continuous at a, the second that it is continuous at b, and the third that it is continuous at every point of (a, b) (5.5.6). $\Diamond$

Exercises

1. State and prove the 'left' analogue of 5.5.5.

2. Suppose $f : \mathrm{S} \to \mathbb{R}$, $g : \mathrm{S} \to \mathbb{R}$ are right continuous at $a \in \mathrm{S}$, and let $c \in \mathbb{R}$. Prove:
(i) $f + g$, fg and cf are right continuous at a.
(ii) If, moreover, $g(a) \neq 0$, then the set $\mathrm{T} = \{x \in \mathrm{S} : g(x) \neq 0\}$ is a right neighborhood of a, and the function $f/g : \mathrm{T} \to \mathbb{R}$ is right continuous at a. {For $x \in \mathrm{T}$, $(f/g)(x)$ is defined to be $f(x)/g(x)$.}

3. Let $a \in \mathrm{N} \subset \mathbb{R}$. The following conditions are equivalent:
(a) N is *not* a right neighborhood of a;
(b) there exists a sequence (x_n) in the complement $\complement \mathrm{N}$ of N such that $x_n > a$ and $x_n \to a$.

4. Call a set $\mathrm{A} \subset \mathbb{R}$ *right-open* if it is a right neighborhood of each of its points (cf. 4.3.4). By convention, $\varnothing$ is right-open. The analogue of 4.3.8 is true:
(i) $\varnothing$, $\mathbb{R}$ are right-open sets;
(ii) if A and B are right-open sets then so is $\mathrm{A} \cap \mathrm{B}$ (cf. 5.5.2);
(iii) if $\mathcal{S}$ is any set of right-open sets in $\mathbb{R}$, then the union $\bigcup \mathcal{S}$ is also right-open.
For example, every interval $[a, b)$ is right-open; it follows that a set $\mathrm{A} \subset \mathbb{R}$ is right-open if and only if it is the union of a set $\mathcal{S}$ of intervals $[a, b)$.

5. If $f : \mathbb{R} \to \mathbb{R}$ the following conditions are equivalent:

(a) f is right continuous at every $a \in \mathbb{R}$;

(b) U open $\Rightarrow$ $f^{-1}(U)$ is right-open (cf. Exercise 4).

6. Let $f : \mathbb{R} \to \mathbb{R}$ be a function such that $f(x+y) = f(x) + f(y)$ for all x, y in $\mathbb{R}$. Suppose that for each $x \in \mathbb{R}$, f is either left continuous at x or right continuous at x. Prove that $f(x) = cx$ for a suitable constant c.

{Hint: Let $c = f(1)$ and note that $f(r) = cr$ for all $r \in \mathbb{Q}$.}

5.6. Composition

5.6.1. The composition of functions—one function followed by another—is familiar from calculus (in connection with the "chain rule" for differentiation). The simplest general setting is as follows (Appendix, A.3.3): we are given functions $f : X \to Y$ and $g : Y \to Z$ (Figure 10),

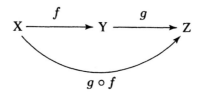

FIGURE 10

where the final set for f is the initial set for g; for $x \in X$, the correspondence

$$x \mapsto f(x) \mapsto g(f(x))$$

produces a function $X \to Z$, called the **composite** of g and f and denoted $g \circ f$ (verbalized "g circle f"). The defining formula for $g \circ f : X \to Z$ is

$$(g \circ f)(x) = g(f(x)) \quad (x \in X).$$

5.6.2. The simplest situation of all is where $X = Y = Z$. For example, if $f : \mathbb{R} \to \mathbb{R}$ and $g : \mathbb{R} \to \mathbb{R}$ are the functions $f(x) = x^2 + 5$ and $g(y) = y^3$ then

$$(g \circ f)(x) = g(f(x)) = (f(x))^3 = (x^2 + 5)^3.$$

Thus $h = g \circ f$ is the function $h(x) = (x^2 + 5)^3$ ($x \in \mathbb{R}$).

5.6.3. Slightly more permissive than the setup of 5.6.1 is the situation where $f : X \to Y$, $g : U \to V$ and $f(X) \subset U$. If $x \in X$ then $f(x) \in f(X) \subset U$, so $g(f(x))$ makes sense; thus $g \circ f : X \to V$ can be defined by the same formula as in 5.6.1.

The relation between composition and continuity is harmonious (to tell the truth, there are a few sour notes in the exercises):

5.6.4. Theorem. *Suppose* $f : S \to \mathbb{R}$, $g : T \to \mathbb{R}$, *where* S *and* T *are subsets of* $\mathbb{R}$ *such that* $f(S) \subset T$, *and let* $a \in S$. *If* f *is continuous at* a, *and* g *is continuous at* $f(a)$, *then* $g \circ f$ *is continuous at* a.

Proof. If $x_n \in S$, $x_n \to a$ then $f(x_n) \to f(a)$ (because f is continuous at a), therefore $g(f(x_n)) \to g(f(a))$ (because g is continuous at $f(a)$), in other words $(g \circ f)(x_n) \to (g \circ f)(a)$. ◊

5.6.5. Corollary. *With notations as in 5.6.4, if* f *and* g *are continuous functions (5.4.1) then so is* $g \circ f$.

Incidentally, it suffices in 5.6.5 that f be continuous and that g be continuous at every point of the range $f(S)$ of f.

5.6.6. Given two functions $f : X \to Y$ and $g : U \to V$ such that the set

$$A = \{x \in X : f(x) \in U\}$$

is nonempty, a function $g \circ f : A \to V$ can be defined by the formula $(g \circ f)(x) = g(f(x))$ $(\forall x \in A)$.

In principle, one can compose any two functions, but the result may be disappointing. For example, if f and g are the functions

$$f : \mathbb{R} \to \mathbb{R}, \quad f(x) = 0 \quad (\forall x \in \mathbb{R})$$
$$g : \mathbb{R} - \{0\} \to \mathbb{R}, \quad g(x) = 1/x \quad (\forall x \neq 0)$$

the formula for $g \circ f$ suggested by the foregoing discussion is $(g \circ f)(x) = g(f(x)) = 1/f(x) = 1/0$. {Not to worry—the domain of $g \circ f$ is the empty set.}

Exercises

1. If $f : \mathbb{R} \to \mathbb{R}$ and $g : \mathbb{R} \to \mathbb{R}$ are continuous functions, we know that $g \circ f : \mathbb{R} \to \mathbb{R}$ is continuous (5.6.5). Give an alternate proof using open sets.

2. (i) With notations as in 5.6.4, prove that if $f : S \to \mathbb{R}$ is right continuous at $a \in S$ and $g : T \to \mathbb{R}$ is continuous at $f(a)$, then $g \circ f$ is right continuous at a.

(ii) True or false (explain): If $f : \mathbb{R} \to \mathbb{R}$ is continuous at a, and $g : \mathbb{R} \to \mathbb{R}$ is right continuous at $f(a)$, then $g \circ f$ is right continuous at a. {Hint: Let $g(x) = [x]$ (2.3.2) and let $f(x) = x$ for x rational, $f(x) = -x$ for x irrational.}

(iii) Give an example of an everywhere discontinuous function $f : \mathbb{R} \to \mathbb{R}$ and a nonconstant, everywhere continuous function $g : \mathbb{R} \to \mathbb{R}$ such that $g \circ f$ is everywhere continuous. {Hint: Consider $g(x) = |x|$.}

CHAPTER 6

Continuous Functions
on an Interval

Intervals are convex (4.1.4) and closed intervals are compact (4.5.4); these special properties of intervals are reflected in special properties of continuous functions defined on them. Applications include the construction of n'th roots, a characterization of injectivity (§6.4) and automatic continuity of inverse functions (§6.5).

6.1. Intermediate Value Theorem

If a continuous real-valued function on a closed interval has opposite signs at the endpoints, then it must be zero somewhere in between:

6.1.1. Lemma. *If $f : [a, b] \to \mathbb{R}$ is a continuous function such that $f(a)f(b) < 0$, then there exists a point $c \in (a, b)$ such that $f(c) = 0$.*

Proof. We can suppose $f(a) > 0$ and $f(b) < 0$ (otherwise consider $-f$). The idea behind the following proof: there are points x in $[a, b]$ (for example, $x = a$) for which $f(x) \geq 0$, and b isn't one of them; the 'last' such point x is a likely candidate for c.

The set $A = \{x \in [a, b] : f(x) \geq 0\}$ is nonempty (because $a \in A$) and bounded. It is also closed: for, if $x_n \in A$ and $x_n \to x$, then $x \in [a, b]$ (4.2.7) and $f(x_n) \to f(x)$ by the continuity of f; since $f(x_n) \geq 0$ for all n, $f(x) \geq 0$ by 3.4.8, (8), thus $x \in A$.

Let c be the largest element of A (proof of 4.5.7). In particular, $f(c) \geq 0$, therefore $c \neq b$, thus $c < b$. If $c < x < b$ then $x \notin A$ (all elements of A are $\leq c$), so $f(x) < 0$. Choose a sequence (x_n) with $c < x_n < b$ and $x_n \to c$; then $f(c) = \lim f(x_n) \leq 0$, consequently $f(c) = 0$. ◇

6.1.2. Theorem. (Intermediate Value Theorem) *If* I *is an interval in* $\mathbb{R}$ *and* $f : I \to \mathbb{R}$ *is continuous, then* $f(I)$ *is also an interval.*

Proof. Assuming $r, s \in f(I)$, $r < s$, it will suffice to show that $[r, s] \subset f(I)$ (4.1.4). Let $r < k < s$; we seek $c \in I$ such that $f(c) = k$. {The message of the theorem: If r and s are values of f, then so is every number between r and s.} By assumption $r = f(a)$ and $s = f(b)$ for suitable points a, b of I; since $r \neq s$, also $a \neq b$. Let J be the closed interval with endpoints a and b (whether $a < b$ or $b < a$ does not interest us); since I is an interval, $J \subset I$. Define $g : J \to \mathbb{R}$ by the formula

$$g(x) = f(x) - k \quad (x \in J).$$

Since f is continuous, so is g; moreover,

$$g(a) = f(a) - k = r - k < 0,$$
$$g(b) = f(b) - k = s - k > 0.$$

By the lemma, there exists a point $c \in J$ such that $g(c) = 0$; thus $c \in I$ and $f(c) - k = 0$, so $k = f(c) \in f(I)$. ◇

6.1.3. Corollary. *With* $f : I \to \mathbb{R}$ *as in the theorem, if* f *is not zero at any point of* I *then either* $f(x) > 0$ $(\forall\, x \in I)$ *or* $f(x) < 0$ $(\forall x \in I)$.

Proof. The alternative is that $f(a) < 0$ and $f(b) > 0$ for suitable points a, b of I; then $0 \in f(I)$ by the theorem, contrary to the hypothesis on f. ◇

6.1.4. Corollary. *If* $f : \mathbb{R} \to \mathbb{R}$ *is continuous and* I *is any interval in* $\mathbb{R}$, *then* $f(I)$ *is also an interval.*

Proof. Apply 6.1.2 to the function $f|I : I \to \mathbb{R}$ (the restriction of f to I). ◇

Applications of the Intermediate Value Theorem (briefly, IVT) are given in the next four sections and in the exercises.

Exercises

1. If p is a polynomial of odd degree, with real coefficients, then p has at least one real root.
{Hint: For example, if $p(x) = 5x^3 - 2x^2 + 3x - 4$ then

$$f(n) = n^3(5 - 2/n + 3/n^2 - 4/n^3) \quad (\forall\, n \in \mathbb{P});$$

the second factor converges to 5 as $n \to \infty$, so f has values > 0. Similarly, consideration of $f(-n)$ shows that f has negative values.}

2. True or false (explain): If I is a bounded interval and $f : I \to \mathbb{R}$ is continuous, then f is bounded.

3. True or false (explain): If $f : \mathbb{R} \to \mathbb{R}$ is continuous and I is an interval in $\mathbb{R}$, then $f^{-1}(I)$ is an interval.

4. Every continuous function $f : [0,1] \to [0,1]$ has a fixed point, that is, there exists a point $c \in [0,1]$ such that $f(c) = c$.
{Hint: Let $g(x) = f(x) - x$ and consider the possibilities for $g(0)$ and $g(1)$.}

*6.2. n'th Roots

The Dedekind cut technique used to construct square roots (2.8.1) can be adapted to higher-order roots (Exercise 1), but the Intermediate Value Theorem provides an efficient shortcut:

6.2.1. Theorem. *If n is a positive integer and $f : [0, +\infty) \to [0, +\infty)$ is the function defined by $f(x) = x^n$, then f is bijective.*

Proof. From 1.2.8 we see that $f(a) = f(b) \Rightarrow a = b$, so f is injective. Write $I = [0, +\infty)$; thus $f : I \to I$ and it remains to show that f is surjective, that is, $f(I) = I$.

Since f is continuous, its range $f(I)$ is an interval by the IVT (6.1.2). From $f(0) = 0$ we have $0 \in f(I)$, and an easy induction argument shows that $f(k) \geq k$ for every positive integer k; it follows that $[0, k] \subset f(I)$ for all $k \in \mathbb{P}$, therefore (Archimedes) $[0, +\infty) \subset f(I)$. Thus $I \subset f(I) \subset I$, whence equality. ◊

6.2.2. Definition. If $x \geq 0$ and n is a positive integer, the unique $y \geq 0$ such that $y^n = x$ (6.2.1) is called the n'**th root** of x, written $y = \sqrt[n]{x}$ (or $y = x^{1/n}$).

6.2.3. Corollary. *If n is an odd positive integer and $g : \mathbb{R} \to \mathbb{R}$ is the function defined by $g(x) = x^n$, then g is bijective.*

Proof. By the IVT, $g(\mathbb{R})$ is an interval, and $g(\mathbb{R})$ contains $[0, +\infty)$ by 6.2.1; but $g(-x) = -g(x)$ (because n is odd), so $g(\mathbb{R})$ also contains $(-\infty, 0]$, thus is equal to $\mathbb{R}$. Injectivity is easily deduced from 6.2.1 since x and $g(x)$ have the same sign. (Alternatively, cite §1.2, Exercise 5.) ◊

6.2.4. Definition. If $x \in \mathbb{R}$ and $n \in \mathbb{P}$ is odd, the unique real number y such that $y^n = x$ (6.2.3) is called the n'**th root** of x, written $\sqrt[n]{x}$ (or $x^{1/n}$); when $x \geq 0$, this is consistent with 6.2.2.

---

*Optional. The logarithmic function provides an alternate path to n'th roots (9.5.15).

6.2.5. *Remark.* If n is even, then $\sqrt[n]{x}$ is defined only for $x \geq 0$; when n is odd, x can be any real number. In either case, $\sqrt[n]{x}$ always has the sign of x.

Exercises

1. Prove 6.2.1 by adapting the Dedekind cut technique of 2.8.1. {Hint: Given $c > 0$ we seek $y > 0$ with $y^n = c$. Let

$$A = \{x \in \mathbb{R} : x \leq 0\} \cup \{x \in \mathbb{R} : x > 0 \text{ and } x^n < c\}$$
$$B = \{x \in \mathbb{R} : x > 0 \text{ and } x^n \geq c\}.$$

Argue that A has no largest element as follows. Suppose $x > 0$, $x^n < c$. The sequence

$$(x + 1/k)^n - x^n \quad (k = 1, 2, 3, \ldots)$$

is null (binomial formula) and $c - x^n > 0$, so ultimately $(x+1/k)^n - x^n < c - x^n$, that is, $(x + 1/k)^n < c.$}

2. If $0 \leq x < y$ then $0 < \sqrt[n]{y} - \sqrt[n]{x} \leq \sqrt[n]{y - x}$. Infer that the functions $x \mapsto \sqrt[n]{x}$ of 6.2.2, 6.2.4 are continuous.

6.3. Continuous Functions on a Closed Interval

A continuous image of a closed interval is a closed interval:

6.3.1. **Theorem.** *If $f : [a, b] \to \mathbb{R}$ is continuous, then the range of f is a closed interval.*

Proof. Write $I = [a, b]$; we know from the IVT (6.1.2) that $f(I)$ is an interval, so we need only show that $f(I)$ is (i) bounded and (ii) a closed set.

(i) The claim is that the set $\{|f(x)| : x \in I\}$ is bounded above. Assume to the contrary. For each positive integer n, choose a point $x_n \in I$ such that $|f(x_n)| > n$. It is clear that no subsequence of $(f(x_n))$ is bounded. However, (x_n) is bounded, so it has a convergent subsequence (Weierstrass–Bolzano), say $x_{n_k} \to x$. Then $x \in I$ (because I is a closed set) and $f(x_{n_k}) \to f(x)$; in particular, $(f(x_{n_k}))$ is bounded, a contradiction.

(ii) Suppose $y_n \in f(I)$, $y_n \to y$; we have to show that $y \in f(I)$. Say $y_n = f(x_n)$, $x_n \in I$; passing to a subsequence, we can suppose $x_n \to x \in \mathbb{R}$. As in the proof of (i), $x \in I$ and $f(x_n) \to f(x)$, that is, $y_n \to f(x)$; but $y_n \to y$, so $y = f(x) \in f(I)$. $\Diamond$

6.3.2. Corollary. (Weierstrass) *If $f : [a,b] \to \mathbb{R}$ is continuous, then f takes on a smallest value and a largest value.*

Proof. By the theorem, $f([a,b]) = [m, M]$ for suitable m and M. Thus, if $m = f(c)$ and $M = f(d)$, then $f(c) \le f(x) \le f(d)$ for all $x \in [a,b]$. $\Diamond$

The continuous function $(0,1] \to \mathbb{R}$ defined by $x \mapsto x$ is > 0 at every point of its domain, but it has values as near to 0 as we like; on a closed interval, that can't happen:

6.3.3. Corollary. *If $f : [a,b] \to \mathbb{R}$ is continuous and $f(x) > 0$ for all $x \in [a,b]$, then there exists an $m > 0$ such that $f(x) \ge m$ for all $x \in [a,b]$.*

Proof. (So to speak, f is 'bounded away from 0'.) With notations as in the preceding proof, we have $f(x) \ge m = f(c) > 0$ for all $x \in [a,b]$. $\Diamond$

6.3.4. Definition. A real-valued function $f : X \to \mathbb{R}$ is said to be **bounded** if its range $f(X)$ is a bounded subset of $\mathbb{R}$, that is, if there exists a real number $M > 0$ such that $|f(x)| \le M$ for all $x \in X$; f is said to be **unbounded** if it is not bounded.

6.3.5. Examples. Every continuous real-valued function on a *closed* interval is bounded (6.3.1). The continuous function $f : (0,1] \to \mathbb{R}$ defined by $f(x) = 1/x$ is unbounded.

Exercises

1. Give a concise proof of 6.3.1 based on the fact that the closed intervals are the compact intervals (§4.5, Exercise 4). {Hint: §5.4, Exercise 7.}

2. The continuous function $[1, +\infty) \to \mathbb{R}$ defined by $x \mapsto 1/x$ has no smallest value, although the interval $[1, +\infty)$ is a closed set. Does this conflict with 6.3.2?

3. Find a bounded continuous function $f : (0,1] \to \mathbb{R}$ such that f cannot be extended to a continuous function $[0,1] \to \mathbb{R}$. {Hint: First contemplate the behavior of $\sin(1/x)$ as $x \to 0$, then simulate the pathological behavior by a simpler function (for example, a 'sawtooth' function).}

4. A polynomial function $p : \mathbb{R} \to \mathbb{R}$ is bounded if and only if it is constant.

6.4. Monotonic Continuous Functions

6.4.1. *Definition.* Let $S \subset \mathbb{R}$ (in the most important examples, S is an interval). A function $f : S \to \mathbb{R}$ is said to be

> **increasing,** if $x < y \Rightarrow f(x) \le f(y)$,
>
> **strictly increasing,** if $x < y \Rightarrow f(x) < f(y)$,
>
> **decreasing,** if $x < y \Rightarrow f(x) \ge f(y)$,
>
> **strictly decreasing,** if $x < y \Rightarrow f(x) > f(y)$,

where it is understood that x and y are in the domain S of f. If f is either increasing or decreasing, it is said to be **monotone**; a function is **strictly monotone** if it is strictly increasing or strictly decreasing.

6.4.2. *Examples.* Let n be a positive integer. The function

$$f : [0, +\infty) \to [0, +\infty), \quad f(x) = x^n$$

is strictly increasing (cf. 1.2.8, 6.2.1); so is the function

$$g : \mathbb{R} \to \mathbb{R}, \quad g(x) = x^{2n-1}$$

(cf. 6.2.3). The function $\mathbb{R} \to \mathbb{R}$ defined by $x \mapsto x^2$ is neither increasing nor decreasing.

6.4.3. *Example.* The function $(0, +\infty) \to (0, +\infty)$ defined by $x \mapsto 1/x$ is strictly decreasing (§1.2, Exercise 1).

6.4.4. *Examples.* Every constant function is increasing (and decreasing), but not strictly. If a function is both increasing and decreasing, then it is a constant function.

*6.4.5. *Examples.* The functions

$$\log : (0, +\infty) \to \mathbb{R}, \quad \exp : \mathbb{R} \to (0, +\infty)$$

are both strictly increasing (here the base is e). These functions are constructed in §9.5.*

If $f : S \to \mathbb{R}$ is strictly monotone, it is obvious that f is injective. Here is a deeper result in the reverse direction:

6.4.6. *Theorem.* *If* $f : [a, b] \to \mathbb{R}$ *is continuous and injective, then* f *is strictly monotone.*

Proof. If $a = b$ there is nothing to prove; assuming $a < b$, $f(a) \ne f(b)$ by injectivity. We can suppose $f(a) < f(b)$ (if not, consider $-f$); let's show that f is then strictly increasing. The heart of the matter is the following *claim*: If $a < x < b$ then $f(a) < f(x) < f(b)$.

Assume to the contrary that $f(x) \leq f(a)$ or $f(x) \geq f(b)$, in other words (injectivity) $f(x) < f(a)$ or $f(x) > f(b)$. In the first case, $f(x) < f(a) < f(b)$, thus $k = f(a)$ is intermediate to the values of $f|[x,b]$ at the endpoints of $[x,b]$; the IVT then yields a point $t \in (x,b)$ with $f(t) = k = f(a)$, contrary to injectivity. In the second case, $f(a) < f(b) < f(x)$, and an application of the IVT to $f|[a,x]$ yields a point $t \in (a,x)$ with $f(t) = f(b)$, again contradicting injectivity.

Assuming now that $a \leq c < d \leq b$, we have to show that $f(c) < f(d)$. If $a = c$ and $d = b$ there is nothing to prove. If $a = c < d < b$ then $f(c) < f(d)$ by the claim, and similarly if $a < c < d = b$. Finally, if $a < c < d < b$ then $f(a) < f(c) < f(b)$ by the claim applied to $a < c < b$; but then $f(c) < f(d) < f(b)$ by the claim applied to $c < d < b$ and the function $f|[c,b]$. ◊

The theorem extends easily to functions on every kind of interval:

6.4.7. Corollary. *If* I *is an interval and* $f : I \to \mathbb{R}$ *is continuous and injective, then* f *is strictly monotone.*

Proof. If I is a singleton there is nothing to prove. Otherwise, let $r, s \in I$ with $r < s$. Since f is injective, $f(r) \neq f(s)$; we can suppose $f(r) < f(s)$ (if not, consider $-f$). We assert that f is strictly increasing. Given $c, d \in I$, $c < d$, we must show that $f(c) < f(d)$. Let $J = [a,b]$ be a closed subinterval of I that contains all four points r, s, c, d; for example,

$$a = \min\{r, c\}, \ b = \max\{s, d\}$$

fill the bill. From the theorem, we know that $f|J$ is either strictly increasing or strictly decreasing; since $f(r) < f(s)$ it must be the former, in particular $f(c) < f(d)$. ◊

Exercises

1. Let $f : \mathbb{R} - \{0\} \to \mathbb{R} - \{0\}$ be the function $f(x) = 1/x$.
(i) If $a < b < 0$ then $f(a) > f(b)$.
(ii) True or false (explain): f is decreasing. {Hint: Sketch the graph.}

2. If $f : S \to \mathbb{R}$ is increasing, then f is injective if and only if it is strictly increasing. A monotone function is injective if and only if it is strictly monotone.

3. Let S and T be subsets of $\mathbb{R}$, $f : S \to T$ a bijection, $f^{-1} : T \to S$ the inverse function. If f is strictly increasing then so is f^{-1}.

4. Let $c \in S \subset \mathbb{R}$. A function $f : S \to \mathbb{R}$ is said to have a *local maximum* at c if there exists a neighborhood V of c in $\mathbb{R}$ such that

$V \subset S$ and $f(x) \leq f(c)$ for all $x \in V$. Local minima are defined similarly (just change the inequality to $f(x) \geq f(c)$).

Prove: If $f : [a, b] \to \mathbb{R}$ is continuous and has no local maximum or local minimum, then f is strictly monotone.

5. Let I be an interval, $f : I \to \mathbb{R}$ an increasing function such that $f(I)$ is an interval. Prove:

(i) If $a, b \in I$, $a \leq b$, then $f([a, b]) = [f(a), f(b)]$. {Sketch of proof: The inclusion $\subset$ is obvious. Conversely, suppose $f(a) < k < f(b)$. Then $k \in [f(a), f(b)] \subset f(I)$, say $k = f(c)$, $c \in I$; from $f(a) < f(c) < f(b)$ infer that $a < c < b$, whence $k = f(c) \in f([a, b])$.}

(ii) f is continuous. {Sketch of proof: Suppose $a_n \to a$ in I; we must show $f(a_n) \to f(a)$. Let $\epsilon > 0$ and assume to the contrary that $|f(a_n) - f(a)| > \epsilon$ frequently. Say $f(a_n) - f(a) > \epsilon$ frequently; passing to a subsequence, we can suppose $f(a_n) > f(a) + \epsilon$ for all n. From $f(a) < f(a) + \epsilon < f(a_1)$ and (i), we have $f(a) + \epsilon = f(c)$ for some c; and $f(a) < f(c) < f(a_n)$ implies $a < c < a_n$ for all n, contrary to $a_n \to a$.}

6. Let n be a positive integer.

(i) The function $[0, +\infty) \to [0, +\infty)$ defined by $x \mapsto \sqrt[n]{x}$ is continuous. {Hint: Exercises 3 and 5.}

(ii) If n is odd, the function $\mathbb{R} \to \mathbb{R}$ defined by $x \mapsto \sqrt[n]{x}$ is continuous.

7. If I and J are intervals and $f : I \to J$ is bijective, the following conditions are equivalent: (a) f is continuous; (b) f is monotone. If the conditions are verified then f^{-1} is also continuous. {Hint: Exercises 3 and 5.}

8. If U is an open set in $\mathbb{R}$ and $f : U \to \mathbb{R}$ is continuous and injective, then $f(U)$ is also an open set. {Hint: Reduce to the case that U is an open interval.}

9. For every interval I, there exists an increasing bijection of I onto one of the intervals $[0, 1]$, $(0, 1)$, $[0, 1)$, $(0, 1]$.

{Hint: The function g of §5.4, Exercise 8, (ii) is strictly increasing.}

6.5. Inverse Function Theorem

Under the right conditions, the inverse is continuous 'free of charge':

6.5.1. Lemma. *If $f : [a, b] \to [c, d]$ is bijective and continuous then the inverse function $f^{-1} : [c, d] \to [a, b]$ is also continuous.*

Proof. Assuming $y_n \to y$ in $[c, d]$, we are to show that $f^{-1}(y_n) \to f^{-1}(y)$. Let $x_n = f^{-1}(y_n)$, $x = f^{-1}(y)$ and assume to the contrary that (x_n) does not converge to x. Then there exists an $\epsilon > 0$ such that $|x_n - x| \geq \epsilon$ frequently; passing to a subsequence, we can suppose that $|x_n - x| \geq \epsilon$ for all n. Since (x_n) is bounded, some subsequence is convergent (Weierstrass–Bolzano), say $x_{n_k} \to t$. Then $t \in [a, b]$ so $f(x_{n_k}) \to f(t)$ by continuity; but $f(x_{n_k}) = y_{n_k} \to y$, so $y = f(t)$, $t = f^{-1}(y) = x$. Thus $x_{n_k} \to x$, contrary to $|x_{n_k} - x| \geq \epsilon$ $(\forall\ k)$. $\Diamond$

6.5.2. Theorem. (Inverse Function Theorem) *Let* I *be an interval in* $\mathbb{R}$, $f : I \to \mathbb{R}$ *continuous and injective; let* $J = f(I)$ *(an interval, by 6.1.2), so that* $f : I \to J$ *is continuous and bijective. Then* $f^{-1} : J \to I$ *is also continuous.*

Proof. From 6.4.7 we know that f is monotone; we can suppose that f is increasing (if not, consider $-f$). Suppose $y_n \to y$ in J; writing $x_n = f^{-1}(y_n)$, $x = f^{-1}(y)$, we have to show that $x_n \to x$.

The set $A = \{y\} \cup \{y_n : n \in \mathbb{P}\}$ is compact (4.5.2, 4.5.5), so it has a smallest element c and a largest element d (4.5.7). Then $A \subset [c, d] \subset J$. Say $c = f(a)$, $d = f(b)$; since f is increasing, $a \leq b$ and $f([a, b]) = [f(a), f(b)] = [c, d]$, so $x_n \to x$ follows from applying the lemma to the restriction $f|[a, b] : [a, b] \to [c, d]$. $\Diamond$

6.5.3. *Example.* If n is a positive integer and $I = [0, +\infty)$, then the function $I \to I$ defined by $x \mapsto \sqrt[n]{x}$ is continuous; for, it is the inverse of a continuous bijection (6.2.1). If n is odd, then the function $\mathbb{R} \to \mathbb{R}$ defined by $x \mapsto \sqrt[n]{x}$ is continuous (cf. 6.2.3).

Exercises

1. Let $f : [a, b] \to [c, d]$ be monotone and bijective.
(i) If f is increasing and (x_n) is any sequence in $[a, b]$, show that

$$\limsup f(x_n) = f(\limsup x_n)$$

and similarly for $\liminf$.
(ii) Deduce from (i) that f is continuous. {Hint: If f is increasing, cite (i) and 3.7.5. If f is decreasing, consider $-f$.}
(iii) Infer from (ii) that f^{-1} is also continuous.

2. The reference to compactness in the proof of 6.5.2 is avoidable (§4.1, Exercise 8).

*6.6. Uniform Continuity

6.6.1. Theorem. *Suppose* $f : [a, b] \to \mathbb{R}$ *is continuous. Given any* $\epsilon > 0$, *there exists a* $\delta > 0$ *such that*

$$x, y \in [a, b], \ |x - y| < \delta \ \Rightarrow \ |f(x) - f(y)| < \epsilon.$$

---

*Omissible (6.6.1 is cited only in §9.7, Exercise 5).

This is not just a restatement of the definition of continuity; there is a subtle difference: to say that $f : S \to \mathbb{R}$ is continuous means that for each $y \in S$ and $\epsilon > 0$ there is a $\delta > 0$ (depending in general on both y and ϵ) such that

$$x \in S, \ |x - y| < \delta \ \Rightarrow \ |f(x) - f(y)| < \epsilon.$$

The message of the theorem: when the domain of f is a closed interval, the choice of δ can be made to depend on ϵ alone; so to speak, δ works 'uniformly well' at all points of the domain.

Proof of 6.6.1. Let $\epsilon > 0$; we seek a $\delta > 0$ for which the stated implication is valid. Assume to the contrary that no such δ exists. In particular, for each $n \in \mathbb{P}$ the choice $\delta = 1/n$ fails to validate the implication, so there is a pair of points x_n, y_n in $[a, b]$ such that

$$|x_n - y_n| < 1/n \ \text{ but } \ |f(x_n) - f(y_n)| \geq \epsilon.$$

For a suitable subsequence, $x_{n_k} \to x \in [a, b]$ (Weierstrass–Bolzano); then $y_{n_k} = x_{n_k} - (x_{n_k} - y_{n_k})$ and $x_{n_k} - y_{n_k} \to 0$ show that also $y_{n_k} \to x$. By continuity,

$$f(x_{n_k}) \to f(x) \ \text{ and } \ f(y_{n_k}) \to f(x),$$

so $f(x_{n_k}) - f(y_{n_k}) \to 0$, contrary to $|f(x_{n_k}) - f(y_{n_k})| \geq \epsilon$. ◊

6.6.2. *Definition.* Let S be a subset of $\mathbb{R}$. A function $f : S \to \mathbb{R}$ is said to be **uniformly continuous** (on S) if, for every $\epsilon > 0$, there exists a $\delta > 0$ such that

$$x, y \in S, \ |x - y| < \delta \ \Rightarrow \ |f(x) - f(y)| < \epsilon.$$

It is obvious that uniform continuity implies continuity, but the converse is false:

6.6.3. *Example.* The function $f : (0, 2] \to \mathbb{R}$ defined by $f(x) = 1/x$ is continuous but not uniformly continuous. This is 'obvious' from the graph of f. {The nearer y is to 0, the steeper the 'slope' of the graph; for a particular ϵ, the nearer y is to 0, the smaller δ will have to be taken.} Here's a formal argument. Assume to the contrary that f is uniformly continuous. In particular, for $\epsilon = 1$ there is a $\delta > 0$ (which we can suppose to be < 1) for which

$$x, y \in (0, 2], \ |x - y| \leq \delta \ \Rightarrow \ |1/x - 1/y| < 1.$$

The last inequality may be written $|x - y| < xy$, so in particular

(∗) $x, y \in (0, 2], \ |x - y| = \delta \ \Rightarrow \ \delta < xy.$

If $y_n = 1/n$ and $x_n = \delta + 1/n$, then $|x_n - y_n| = \delta < x_n y_n$ by (*); since $x_n y_n \to 0$, this is absurd.

Exercises

1. If A is a compact subset of $\mathbb{R}$ (4.5.5) then every continuous function $f : A \to \mathbb{R}$ is uniformly continuous. {Hint: Adapt the proof of 6.6.1 (cf. §4.5, Exercise 1).}

2. Let $S \subset \mathbb{R}$, $f : S \to \mathbb{R}$. Consider the statements:
(i) $(\forall\, y \in S)\ (\forall\, \epsilon > 0)\ \exists\, \delta > 0 \ni$

$$x \in S,\ |x - y| < \delta \ \Rightarrow\ |f(x) - f(y)| < \epsilon.$$

(ii) $(\forall\, \epsilon > 0)\ \exists\, \delta > 0 \ni (\forall\, y \in S)$

$$x \in S,\ |x - y| < \delta \ \Rightarrow\ |f(x) - f(y)| < \epsilon.$$

One of these says that f is continuous, the other that f is uniformly continuous. Which is which?

3. The function $f : \mathbb{R} \to \mathbb{R}$ defined by $f(x) = 1/(1 + x^2)$ is uniformly continuous. {Hint: Show first that $|f(x) - f(y)| \le |x - y|$ for all x and y.}

4. If $f : S \to \mathbb{R}$ is uniformly continuous and (x_n) is a Cauchy sequence in S, then $(f(x_n))$ is also a Cauchy sequence.

5. (i) If $f : S \to \mathbb{R}$ and $g : S \to \mathbb{R}$ are uniformly continuous then so are $f + g$, $|f|$, $\sup(f, g)$ and $\inf(f, g)$ (see 5.3.1 and §5.3, Exercise 1 for the notations). What about fg? (Be suspicious!)
(ii) If $f : S \to \mathbb{R}$ and $g : T \to \mathbb{R}$ are uniformly continuous and $f(S) \subset T$, then the composite function $g \circ f : S \to \mathbb{R}$ is uniformly continuous.

6. If S is a bounded subset of $\mathbb{R}$ and $f : S \to \mathbb{R}$ is uniformly continuous, then f is bounded. {Hint: Exercise 4.}

7. Let (X, d), (Y, D) be metric spaces. A function $f : X \to Y$ is said to be *uniformly continuous* (for the given metrics) if, for every $\epsilon > 0$, there exists a $\delta > 0$ such that

$$x, x' \in X,\ d(x, x') < \delta \ \Rightarrow\ D\big(f(x), f(x')\big) < \epsilon.$$

Prove:
(i) The analogue of Exercise 4 holds for a function $f : X \to Y$.
(ii) If X is compact (cf. §4.5, Exercise 6) and $f : X \to Y$ is continuous, then f is uniformly continuous. {Hint: Adapt the proof of 6.6.1.}

8. Let $f : [0, +\infty) \to [0, +\infty)$ be the continuous bijection $f(x) = \sqrt{x}$ (cf. 6.2.1, 6.5.3). Prove 'directly' (using inequalities and the formula for f) that f is uniformly continuous.

{Hint: Extrapolate from $x = 0$, where the tangent line is steepest.}

9. If $S \subset \mathbb{R}$ and $f : S \to \mathbb{R}$, the following conditions are equivalent: (a) f is uniformly continuous; (b) $x_n, y_n \in S$, $x_n - y_n \to 0 \Rightarrow f(x_n) - f(y_n) \to 0$.

CHAPTER 7

Limits of Functions

Our main reason for taking up limits is to prepare the way for derivatives in the next chapter. More generally, limits provide a framework for discussing 'indeterminate forms' ($0/0$, 0^0, $0 \cdot \infty$, etc.)[1]. For example, in calculating the derivative of a function f at a point c, we look at the difference quotient

$$\frac{f(x) - f(c)}{x - c}$$

and decide what happens as x *approaches* c while *avoiding* c; this amounts to evaluating an 'indeterminate form' of type $0/0$.

The integral can also be regarded as a kind of limit (§9.8); this is not the most intuitive way to *define* the integral, but it's nice to know there's a single concept that unifies the fundamental processes of calculus.

7.1. Deleted Neighborhoods

The idea of 'deleted neighborhood' of a point c is to permit a variable x to approach c without ever having to be actually equal to c. The formal definition is as follows:

7.1.1. Definition. Let S be a subset of $\mathbb{R}$, c a real number: $S \subset \mathbb{R}$, $c \in \mathbb{R}$. (We do not assume that c belongs to S, but it *might*.) We say that

S is a *deleted right neighborhood* (DRN) of c if there is an $r > 0$ such that $(c, c+r) \subset S$ (that is, $c < x < c+r \Rightarrow x \in S$);

---

[1] Expressions that are, at first glance, nonsense.

S is a *deleted left neighborhood* (DLN) of c if there is an $r > 0$ such that $(c - r, c) \subset S$ (that is, $c - r < x < c \Rightarrow x \in S$);

S is a *deleted neighborhood* (DN) of c if there is an $r > 0$ such that $(c - r, c) \cup (c, c + r) \subset S$ (that is, $0 < |x - c| < r \Rightarrow x \in S$).

7.1.2. Remarks. (i) S is a deleted neighborhood of c if and only if it is both a deleted left neighborhood and a deleted right neighborhood of c.

(ii) If S is a deleted neighborhood of c and if $S \subset T \subset \mathbb{R}$, then T is also a deleted neighborhood of c; in particular, every neighborhood of c is a deleted neighborhood of c. Similarly for DRN's and DLN's.

(iii) If S and T are DRN's of c, then so is $S \cap T$.

(iv) S is a DRN of c if and only if $S \cup \{c\}$ is a right neighborhood of c (cf. 5.5.1). Similarly for DLN's and DN's.

7.1.3. Example. If $a < b$ then the open interval (a, b) is a DRN of a, a DLN of b, and a neighborhood of every internal point. The same is true of the intervals $[a, b]$, $(a, b]$ and $[a, b)$.

7.1.4. Example. If $f : [a, b] \to \mathbb{R}$, $a < b$, and if $c \in [a, b]$, then the set $[a, b] - \{c\}$ is a DN of c if $c \in (a, b)$; a DRN of c if $c = a$; and a DLN of c if $c = b$. The function $g : [a, b] - \{c\} \to \mathbb{R}$ defined by

$$g(x) = \frac{f(x) - f(c)}{x - c}$$

is familiar from elementary calculus. (See also Chapter 8.)

7.1.5. Remarks on the terminology. If it is disconcerting that a "deleted neighborhood" of c might contain c, reflect on the fact that we also permit a "neighborhood" of c to contain points that are far away from c. For neighborhoods, the points far from c are ignored; for deleted neighborhoods, the presence of c—if it is present—is likewise ignored. The alternatives to "deleted neighborhood" (quasi-neighborhood, etc.) are no more satisfactory; the source of our discomfort is the decision that, whatever we call the sets described in 7.1.1, the ordinary neighborhoods of c are to be included among them.

Exercises

1. True or false (explain):

(i) If S and T are deleted neighborhoods of c, then so is $S \cap T$.

(ii) If S is a DN and T is a DRN of c, then $S \cap T$ is a DRN of c.

(iii) If S is a DLN and T is a DRN of c, then $S \cap T$ is a DN of c; same question for $S \cup T$.

2. If T is a neighborhood of c, then $T - \{c\}$ is a DN of c. State the 'right' and 'left' versions.

3. Deleted neighborhoods can also be defined in metric spaces (call S a deleted neighborhood of c if $S \cup \{c\}$ is a neighborhood of c), but the concept is useless if $\{c\}$ is an open set (such points c are called *isolated* points); in particular, the concept is useless for discrete metric spaces (contemplate the statement of Exercise 2 with $T = \{c\}$).

7.2. Limits

Continuity of a function f at a point a means, informally, that $f(x)$ approaches $f(a)$ as x approaches a. In the theory of limits, f is permitted to be undefined at a provided that, as x approaches a, $f(x)$ approaches *something*:

7.2.1. Definition. Let $f : S \to \mathbb{R}$, where S is a deleted neighborhood of $c \in \mathbb{R}$. We say that f **has a limit** at c if there exists a real number L such that

$$\left. \begin{array}{c} x_n \in S \\ x_n \neq c \\ x_n \to c \end{array} \right\} \Rightarrow f(x_n) \to L.$$

Such a number L is unique (3.4.1); it is called the **limit** of f at c, written

$$\lim_{x \to c} f(x) = L,$$

or, for emphasis,

$$\lim_{x \to c, \, x \neq c} f(x) = L.$$

The statement that f has a limit at c is also expressed by saying that '$\lim_{x \to c} f(x)$ exists'.

7.2.2. Example. If S is a neighborhood of c and $f : S \to \mathbb{R}$ is continuous at c, then f has limit $f(c)$ at c:

$$\lim_{x \to c} f(x) = f(c).$$

Indeed, $f(x_n) \to f(c)$ for *every* sequence (x_n) in S such that $x_n \to c$ (and in particular for those with $x_n \neq c$ for all n).

7.2.3. Examples. If $f : \mathbb{R} - \{3\} \to \mathbb{R}$ is the function defined by $f(x) = x^2$ for $x \neq 3$, then $\lim_{x \to 3} f(x) = 9$. The same is true for the function $f : \mathbb{R} \to \mathbb{R}$ defined by

$$f(x) = \begin{cases} 2x + 3 & \text{for } x < 3 \\ x^2 & \text{for } x > 3 \\ 1 & \text{for } x = 3 \end{cases}$$

(draw a picture!), and for the function $f : \mathbb{R} \to \mathbb{R}$ defined by

$$f(x) = \begin{cases} 2x + 3 & \text{for } x \text{ rational} \\ x^2 & \text{for } x \text{ irrational}. \end{cases}$$

'One-sided limits' are defined by modifying 7.2.1 in the obvious way:

7.2.4. Definition. Let $f : S \to \mathbb{R}$, where S is a deleted right neighborhood of $c \in \mathbb{R}$. We say that f **has a right limit** at c if there exists a real number L such that

$$\left.\begin{array}{r} x_n \in S \\ x_n > c \\ x_n \to c \end{array}\right\} \Rightarrow f(x_n) \to L.$$

Such a number L is unique and is called the **right limit** of f at c, written

$$\lim_{x \to c,\, x > c} f(x) = L$$

or

$$\lim_{x \to c^+} f(x) = L,$$

or, concisely, $f(c+) = L$. The statement that f has a right limit at c is also expressed by saying that '$f(c+)$ exists'. Left limits (when they exist) are defined similarly: S is assumed to be a DLN of c and we require $x_n < c$; the symbols

$$\lim_{x \to c,\, x < c} f(x), \quad \lim_{x \to c^-} f(x), \quad f(c-)$$

denote the left limit of f at c (when it exists).

 7.2.5. Example. If S is a right neighborhood of c and $f : S \to \mathbb{R}$ is right continuous at c, then f has right limit $f(c)$ at c (adapt the argument of 7.2.2); similarly with "right" replaced by "left".

 7.2.6. Example. If $f : \mathbb{R} \to \mathbb{R}$ is defined by

$$f(x) = \begin{cases} 1 & \text{for } x < 0 \\ 2 & \text{for } x = 0 \\ 3 & \text{for } x > 0 \end{cases}$$

then, at the point 0, f has left limit 1 and right limit 3; f does not have a limit at 0 (applying f to the sequence $x_n = (-1)^n/n$ produces a divergent sequence).

 7.2.7. Theorem. *Let* $f : S \to \mathbb{R}$, *where* $S \subset \mathbb{R}$, *and let* $c \in \mathbb{R}$. *The following conditions are equivalent:*

(a) f *has a limit at* c;

(b) $f(c-)$ *and* $f(c+)$ *exist and are equal.*

When f *has a limit* L *at* c, *necessarily* $L = f(c-) = f(c+)$.

Proof. (a) $\Rightarrow$ (b): By assumption, S is a deleted neighborhood of c, so it is also a DLN and a DRN. If f has limit L at c, then $f(x_n) \to L$ for every sequence in S with $x_n \to c$ and $x_n \neq c$; this is true in particular when $x_n < c$ for all n and when $x_n > c$ for all n, thus $f(c-)$ and $f(c+)$ exist and are equal to L.

(b) $\Rightarrow$ (a): By assumption, S is a DLN and a DRN of c, so it is a DN of c. Write L for the common value of $f(c-)$ and $f(c+)$. If (x_n) is a sequence in S with $x_n \to c$ and $x_n \neq c$ for all n, then either (i) $x_n < c$ ultimately, or (ii) $x_n > c$ ultimately, or (iii) $x_n < c$ frequently and $x_n > c$ frequently. In cases (i) and (ii) it is clear that $f(x_n) \to L$. In case (iii), let (x_{n_k}) be the subsequence with $x_{n_k} < c$ and (x_{m_j}) the subsequence with $x_{m_j} > c$; then $f(x_{n_k}) \to f(c-) = L$ and $f(x_{m_j}) \to f(c+) = L$, whence $f(x_n) \to L$. $\Diamond$

Here's a useful theorem on the *existence* of one-sided limits:

7.2.8. Theorem. *If* $f : (a, b) \to \mathbb{R}$ *is a bounded monotone function then* f *has a right limit at every point of* $[a, b)$ *and a left limit at every point of* $(a, b]$.

Proof. It suffices, for example, to show that $f(a+)$ exists.* We can suppose f is increasing (if not, consider $-f$). Let $L = \inf\{f(x) : a < x < b\}$ (recall that f is bounded); we assert that f has right limit L at a. Assuming $a < x_n < b$ and $x_n \to a$, we have to show that $f(x_n) \to L$.

Let $\epsilon > 0$. By the definition of L (as a greatest lower bound) there exists $c \in (a, b)$ such that $L \leq f(c) < L + \epsilon$. Ultimately $a < x_n < c$, therefore $L \leq f(x_n) \leq f(c) < L + \epsilon$, so $|f(x_n) - L| < \epsilon$. We have shown that for every $\epsilon > 0$, $|f(x_n) - L| < \epsilon$ ultimately; in other words, $f(x_n) \to L$. $\Diamond$

7.2.9. Remark. A function $f : [a, b] \to \mathbb{R}$ is said to be **regulated** if f has a right limit at every point of $[a, b)$ and a left limit at every point of $(a, b]$. It follows from 7.2.8 that every monotone function $f : [a, b] \to \mathbb{R}$ is regulated.

Exercises

1. Let $f : S \to \mathbb{R}$, where S is a deleted neighborhood of $c \in \mathbb{R}$. The following conditions are equivalent:

---

*A dual argument shows that $f(b-)$ exists, and if $a < c < b$ then the arguments can be repeated on the restrictions of f to the subintervals (c, b) and (a, c).

(a) f has a limit $L \in \mathbb{R}$ at c;

(b) $x_n \in S$, $x_n \neq c$, $x_n \to c$ $\Rightarrow$ the sequence $(f(x_n))$ is convergent.
(Cf. §5.2, Exercise 4.)

Similarly for left limits and right limits.

2. Let $f : S \to \mathbb{R}$, where S is a deleted right neighborhood of $c \in \mathbb{R}$, and let $L \in \mathbb{R}$. The following conditions are equivalent:

(a) $f(c+) = L$;

(b) $f(x_n) \to L$ for every sequence (x_n) in S such that $x_n > c$ and $x_n \downarrow c$.

{Hint: If $x_n > c$, $x_n \to c$ and if $|f(x_n) - L| \geq \epsilon > 0$ frequently, then the inequality $|f(x_n) - L| \geq \epsilon$ holds for a decreasing subsequence of (x_n) (cf. 3.5.8).}

3. If f has a limit $L \neq 0$ (notations as in 7.2.1) then there exists an $r > 0$ such that $0 < |x - c| < r$ $\Rightarrow$ $x \in S$ and $f(x) \neq 0$. {Hint: Assume to the contrary that $r = 1/n$ 'fails' for every positive integer n.}

4. If $f : [a, b] \to \mathbb{R}$ is a regulated function (7.2.9) then f is bounded. {Hint: Assume to the contrary and apply the Weierstrass–Bolzano theorem.}

5. Suppose $f : [a, b) \to \mathbb{R}$ has a right limit at every point of $[a, b)$. Define $g : [a, b) \to \mathbb{R}$ by $g(x) = f(x+)$ for all $x \in [a, b)$. Prove: g is right continuous at every point of $[a, b)$. {Hint: If $a \leq x < x_n < b$ and $x_n \to x$, choose y_n so that $x_n < y_n < b$, $|x_n - y_n| < 1/n$ and $|f(y_n) - g(x_n)| < 1/n$.}

7.3. Limits and Continuity

For a function defined on a *neighborhood* of a point, continuity at the point means the same thing as having a limit equal to the functional value:

7.3.1. Theorem. *Let $f : S \to \mathbb{R}$, where S is a neighborhood of $c \in \mathbb{R}$. The following conditions on f are equivalent:*

(a) *f is continuous at c;*

(b) $\exists \lim_{x \to c} f(x) = f(c)$.

Proof. (a) $\Rightarrow$ (b): This is 7.2.2.

(b) $\Rightarrow$ (a): Assuming $x_n \in S$, $x_n \to c$, we have to show that $f(x_n) \to f(c)$. This is obvious if $x_n = c$ ultimately, and if $x_n \neq c$ ultimately then it is immediate from (b). The remaining case, that $x_n = c$ frequently and $x_n \neq c$ frequently, follows from applying the preceding two cases to the appropriate subsequences (cf. the proof of 7.2.7). $\Diamond$

7.3.2. *Remarks.* There are 'right' and 'left' versions of 7.3.1. For example, if S is a right neighborhood of c and $f : S \to \mathbb{R}$, then f is right continuous at c if and only if $\exists\, f(c+) = f(c)$. (In the proof of 7.3.1, replace $x_n \neq c$ by $x_n > c$.)

7.3.3. Corollary. *Let* S *be a deleted neighborhood of* $c \in \mathbb{R}$, $f : S \to \mathbb{R}$. *The following conditions on* f *are equivalent:*

(a) f *has a limit at* c;

(b) *there exists a function* $F : S \cup \{c\} \to \mathbb{R}$ *such that* F *is continuous at* c *and* $F(x) = f(x)$ *for all* $x \in S - \{c\}$.

Necessarily $F(c) = \lim_{x \to c} f(x)$.

Proof. Recall that c may or may not belong to S; if $c \in S$ then $S \cup \{c\} = S$, and if $c \notin S$ then $S - \{c\} = S$.

(a) $\Rightarrow$ (b): Say f has limit $L \in \mathbb{R}$ at c. Define $F : S \cup \{c\} \to \mathbb{R}$ as follows:

$$F(x) = \begin{cases} L & \text{if } x = c \\ f(x) & \text{if } x \in S - \{c\}. \end{cases}$$

{If $c \notin S$ we are *extending* f to $S \cup \{c\}$; if $c \in S$ and $f(c) \neq L$, we are *redefining* f at c; if $c \in S$ and $f(c) = L$, nothing has happened, that is, $S \cup \{c\} = S$ and $F = f$.}

If $x_n \in S \cup \{c\}$, $x_n \neq c$, $x_n \to c$, then $x_n \in S - \{c\}$ and $F(x_n) = f(x_n) \to L = F(c)$; by 7.3.1, F is continuous at c.

(b) $\Rightarrow$ (a): Assume F has the properties in (b). If $x_n \in S$, $x_n \neq c$, $x_n \to c$, then $f(x_n) = F(x_n) \to F(c)$ by the continuity of F at c; this shows that f has limit $F(c)$ at c. $\Diamond$

7.3.4. *Remarks.* There are 'one-sided' versions of 7.3.3. For example, if S is a deleted right neighborhood of c and $f : S \to \mathbb{R}$, then f has a right limit at c if and only if there exists a function $F : S \cup \{c\} \to \mathbb{R}$ such that F is right continuous at c and $F(x) = f(x)$ for all $x \in S - \{c\}$; necessarily $F(c) = f(c+)$.

7.3.5. Corollary. *Let* $f : [a, b] \to \mathbb{R}$, $a < b$. *The following conditions are equivalent:*

(a) f *is continuous on* $[a, b]$;

(b) $\exists\, f(a+) = f(a)$, $\exists\, f(b-) = f(b)$ *and, for every* $c \in (a, b)$, $\exists\, \lim_{x \to c} f(x) = f(c)$.

Proof. Condition (b) says that f is right continuous at a, left continuous at b, and continuous at every internal point c, in other words, f is continuous at every point of $[a, b]$; condition (a) says the same thing (5.5.7). $\Diamond$

Exercises

1. Let $f : (a, b) \to \mathbb{R}$ be bounded and monotone. Prove: f can be extended to a function $F : [a, b] \to \mathbb{R}$ that is right continuous at a and left continuous at b. {Hint: 7.2.8.}

2. Write out in full the proofs sketched in 7.3.2 and 7.3.4.

7.4. ϵ, δ **Characterization of Limits**

Limits were defined in 7.2.1 by means of sequences; we can also say it with ϵ's:

7.4.1. Theorem. *Let* $f : S \to \mathbb{R}$, *where* S *is a deleted neighborhood of* $c \in \mathbb{R}$, *and let* $L \in \mathbb{R}$. *The following conditions are equivalent:*
(a) $\exists \lim_{x \to c} f(x) = L$;
(b) *for every* $\epsilon > 0$ *there exists a* $\delta > 0$ *such that*

$$x \in S, \ 0 < |x - c| < \delta \ \Rightarrow \ |f(x) - L| < \epsilon.$$

Proof. Note that there does exist a $\delta > 0$ such that $0 < |x - c| < \delta \Rightarrow x \in S$, in particular $f(x)$ is defined for such x; so the problem in (b) is to assure that in addition $|f(x) - L| < \epsilon$.

Let $F : S \cup \{c\} \to \mathbb{R}$ be the function such that $F(c) = L$ and $F(x) = f(x)$ for all $x \in S - \{c\}$. Condition (b) then says that for every $\epsilon > 0$ there exists a $\delta > 0$ such that

$$x \in S, \ 0 < |x - c| < \delta \ \Rightarrow \ |F(x) - F(c)| < \epsilon,$$

equivalently,

$$x \in S \cup \{c\}, \ |x - c| < \delta \ \Rightarrow \ |F(x) - F(c)| < \epsilon$$

(no harm in letting $x = c$); thus condition (b) is equivalent to the continuity of F at c (5.2.6), which is in turn equivalent to (a) by 7.3.3. $\Diamond$

7.4.2. Remarks. There are one-sided versions of 7.4.1. For example, let $f : S \to \mathbb{R}$, where S is a deleted right neighborhood of $c \in \mathbb{R}$, and let $L \in \mathbb{R}$. In order that f have a right limit at c equal to L, it is necessary and sufficient that for every $\epsilon > 0$ there exist a $\delta > 0$ such that

$$x \in S, \ c < x < c + \delta \ \Rightarrow \ |f(x) - L| < \epsilon.$$

Exercises

1. Let $f : S \to \mathbb{R}$, where S is a deleted neighborhood of $c \in \mathbb{R}$, and let $L \in \mathbb{R}$. The following conditions are equivalent:
(a) $\exists \lim_{x \to c} f(x) = L$;
(b) for every neighborhood V of L in $\mathbb{R}$, $f^{-1}(V)$ is a deleted neighborhood of c. (Cf. 5.2.7.)
Similarly for one-sided limits.

2. Write out in full the proofs of the assertions in 7.4.2.

7.5. Algebra of Limits

The 'algebra of continuity' (§5.3) translates, via 7.3.3, into an 'algebra of limits':

7.5.1. Theorem. *Let* S *be a deleted neighborhood of* $c \in \mathbb{R}$, *and suppose* $f : S \to \mathbb{R}$, $g : S \to \mathbb{R}$ *have limits at* c, *say*

$$\lim_{x \to c} f(x) = L , \quad \lim_{x \to c} g(x) = M .$$

Then the functions $f + g$, fg *and* af ($a \in \mathbb{R}$) *also have limits at* c, *and*

$$\lim_{x \to c} (f + g)(x) = L + M ,$$

$$\lim_{x \to c} (af)(x) = aL ,$$

$$\lim_{x \to c} (fg)(x) = LM .$$

If, moreover, $M \neq 0$, *then* f/g *is defined on a deleted neighborhood of* c *and*

$$\lim_{x \to c} (f/g)(x) = L/M .$$

Proof. Let $F : S \cup \{c\} \to \mathbb{R}$, $G : S \cup \{c\} \to \mathbb{R}$ be the functions such that

$$F(c) = L , \quad F(x) = f(x) \text{ for } x \in S - \{c\} ,$$
$$G(c) = M , \quad G(x) = g(x) \text{ for } x \in S - \{c\} .$$

By 7.3.3, F and G are continuous at c, therefore so is $F + G$ (5.3.1); moreover,

$$(F + G)(c) = F(c) + G(c) = L + M$$

and $(F + G)(x) = f(x) + g(x) = (f + g)(x)$ for $x \in S - \{c\}$, therefore (7.3.3) $f + g$ has a limit at c equal to $L + M$. The proofs for af and fg are similar.

Finally, suppose $M \neq 0$. With $\epsilon = \frac{1}{2}|M|$ in 7.4.1, choose $\delta > 0$ so that

$$0 < |x - c| < \delta \quad \Rightarrow \quad x \in S \text{ and } |g(x) - M| < \frac{1}{2}|M| .$$

In particular, $0 < |x - c| < \delta \Rightarrow g(x) \neq 0$; restricting the functions f and g to the deleted neighborhood $(c - \delta, c) \cup (c, c + \delta)$ of c, we can suppose that g is never 0 on S. Then F/G is continuous at c (5.3.3) and

$$(F/G)(x) = f(x)/g(x) = (f/g)(x) \text{ for } x \in S - \{c\} ,$$

so f/g has limit $F(c)/G(c) = L/M$ at c (7.3.3). $\Diamond$

7.5.2. *Remarks.* There are 'one-sided' versions of 7.5.1. For example, if S is a deleted right neighborhood of c and the functions $f : S \to \mathbb{R}$, $g : S \to \mathbb{R}$ have right limits at c, then the functions $f + g$, af $(a \in \mathbb{R})$ and fg have right limits at c, and

$$(f + g)(c+) = f(c+) + g(c+),$$
$$(af)(c+) = af(c+),$$
$$(fg)(c+) = f(c+)g(c+).$$

If, moreover, $g(c+) \neq 0$ then f/g is defined on a deleted right neighborhood of c and $(f/g)(c+) = f(c+)/g(c+)$.

Exercises

1. Prove 7.5.1 directly from the definition of limit, using sequences (7.2.1).

2. Write out a detailed proof of the assertions in 7.5.2.

3. Let $f : S \to \mathbb{R}$, $g : T \to \mathbb{R}$ and suppose that

$$\lim_{x \to c} f(x) = L, \quad \lim_{y \to L} g(y) = M$$

(in particular, S is a deleted neighborhood of c, and T is a deleted neighborhood of L). Assume there exists an $r > 0$ such that

$$0 < |x - c| < r \quad \Rightarrow \quad x \in S \text{ and } f(x) \neq L.$$

Prove: $g \circ f$ is defined on a deleted neighborhood of c, and

$$\lim_{x \to c} (g \circ f)(x) = M.$$

4. In Exercise 3, if T is a neighborhood of L and g is continuous at L, then the assumption about $f(x) \neq L$ can be dropped.

5. In Exercise 3, with the same assumptions on g, assume only that S is a deleted right neighborhood of c, $f(c+) = L$, and there is an $r > 0$ such that
$$c < x < c + r \quad \Rightarrow \quad x \in S \text{ and } f(x) \neq L.$$

Then $g \circ f$ is defined on a deleted right neighborhood of c and $(g \circ f)(c+) = M$.

6. Suppose
$$\lim_{x \to c} f(x) = L, \quad \lim_{y \to L} g(y) = M$$

and suppose that the limit

$$\lim_{x \to c} (g \circ f)(x)$$

exists (it might not!). Show that the limit of $g \circ f$ at c must equal either M or $g(L)$ (or both).[1]

{Hint: Choose a sequence (x_n) such that $(g \circ f)(x_n)$ is defined, $x_n \neq c$, and $x_n \to c$; either $f(x_n) \neq L$ ultimately, or $f(x_n) = L$ frequently.}

[1] P. Ramankutty and M. K. Vamanamurthy [*American Mathematical Monthly* **82** (1975), 63].

CHAPTER 8

Derivatives

The theme of the chapter: studying a function by studying its rate of change. Although most of the concepts of the chapter are familiar from elementary calculus, some of the refinements of the 'mean value theorem' proved in the last section will probably be new to the reader; these are for application in the chapter on integration.

8.1. Differentiability

8.1.1. *Definition.* Let S be a subset of $\mathbb{R}$, f a real-valued function defined on S, c a point of S; that is, $c \in S \subset \mathbb{R}$ and $f : S \to \mathbb{R}$. Let $g : S - \{c\} \to \mathbb{R}$ be the function defined by the formula

$$g(x) = \frac{f(x) - f(c)}{x - c}$$

(called a 'difference-quotient' function associated with f). We say that

f is **differentiable at** c if S is a neighborhood of c and g has a limit at c;

f is **right differentiable at** c if S is a right neighborhood of c and g has a right limit at c;

f is **left differentiable at** c if S is a left neighborhood of c and g has a left limit at c.

When they exist, these limits are called the **derivative, right deriva-**

tive and **left derivative** of f at c, written

$$f'(c) = \lim_{x \to c} \frac{f(x) - f(c)}{x - c},$$

$$f'_r(c) = \lim_{x \to c^+} \frac{f(x) - f(c)}{x - c},$$

$$f'_l(c) = \lim_{x \to c^-} \frac{f(x) - f(c)}{x - c}.$$

8.1.2. Theorem. *Let $f : S \to \mathbb{R}$, where S is a neighborhood of $c \in \mathbb{R}$. The following conditions on f are equivalent:*
(a) *f is differentiable at c;*
(b) *f is both left and right differentiable at c, and $f'_l(c) = f'_r(c)$.*
For such a function f, necessarily $f'(c) = f'_l(c) = f'_r(c)$.

Proof. This is immediate from 7.2.7 and the definitions. ◇

Just as for general limits, there are sequential and ϵ, δ criteria for differentiability; for example,

8.1.3. Theorem. *Let $f : S \to \mathbb{R}$, where S is a neighborhood of $c \in \mathbb{R}$, and let $L \in \mathbb{R}$. The following conditions on f are equivalent:*
(a) *f is differentiable at c, with derivative L;*
(b) *for every $\epsilon > 0$ there exists a $\delta > 0$ such that if $x \in S$ and $0 < |x - c| < \delta$ then*

$$|f(x) - f(c) - L(x - c)| \le \epsilon|x - c|.$$

Proof. The last inequality in (b) may be written $|g(x) - L| \le \epsilon$, where g is the difference-quotient function (8.1.1), thus the theorem is immediate from 7.4.1. ◇

8.1.4. *Remarks.* There are 'one-sided' versions of 8.1.3. For example, in the criterion for right differentiability, S is a right neighborhood of c and the condition on x in (b) is $c < x < c + \delta$.
Another useful criterion:

8.1.5. Theorem. *Let $f : S \to \mathbb{R}$, where S is a neighborhood of c. The following conditions on f are equivalent:*
(a) *f is differentiable at c;*
(b) *there exists a function $A : S \to \mathbb{R}$ such that A is continuous at c and*

$$f(x) - f(c) = A(x)(x - c) \quad for \ all \ x \in S.$$

A function A satisfying the conditions in (b) is unique, and $f'(c) = A(c)$.

Proof. The equation in condition (b) is trivially satisfied for $x = c$, so the condition means that, in the notation of 8.1.1, there exists a function

$A : S \to \mathbb{R}$ such that A is continuous at c and $A(x) = g(x)$ for all $x \in S - \{c\}$; this is in turn equivalent, by 7.3.3, to the existence of a limit for g at c—in other words, to condition (a)—and the limit is necessarily equal to $A(c)$, that is, $f'(c) = A(c)$. $\Diamond$

8.1.6. *Remarks.* There are one-sided versions of 8.1.5. For example, in the criterion for right differentiability, S is a right neighborhood of c and A is required to be right continuous at c.

8.1.7. **Corollary.** *If* $f : S \to \mathbb{R}$ *is differentiable at* c *then* f *is continuous at* c.

Proof. In the notations of 8.1.5,

$$f(x) = f(c) + A(x)(x - c) \quad \text{for all } x \in S,$$

so f is continuous at c by 5.3.1. $\Diamond$

8.1.8. *Example.* The function $f : \mathbb{R} - \{0\} \to \mathbb{R}$ defined by $f(x) = 1/x$ is differentiable at every $c \in \mathbb{R} - \{0\}$, with $f'(c) = -1/c^2$. {Apply 8.1.5 to the identity

$$1/x - 1/c = (-1/cx)(x - c),$$

citing the continuity of the function $A(x) = -1/cx$ at c (5.3.3).}

Exercises

1. Write out in full the proofs of the assertions sketched in 8.1.4 and 8.1.6.

2. Verify that the function $f : \mathbb{R} \to \mathbb{R}$ defined by $f(x) = |x|$ is both left and right differentiable at 0 but is not differentiable at 0.

3. Prove: If $f : S \to \mathbb{R}$ is right differentiable at $c \in \mathbb{R}$ then f is right continuous at c.

4. Let $f : \mathbb{R} \to \mathbb{R}$ be defined by

$$f(x) = \begin{cases} x & \text{for } x \text{ rational} \\ 0 & \text{for } x \text{ irrational}. \end{cases}$$

Verify that f is continuous at 0 but has neither a right derivative nor a left derivative at 0.

5. Let S be a neighborhood of $c \in \mathbb{R}$ and suppose $f : S \to \mathbb{R}$ is differentiable at c. Let (a_n), (b_n) be sequences in S such that

$$a_n \neq c, \quad b_n \neq c, \quad a_n \neq b_n, \quad a_n \to c, \quad b_n \to c.$$

(i) If $a_n < c < b_n$ for all n, prove that

$$\frac{f(b_n) - f(a_n)}{b_n - a_n} \to f'(c).$$

{Hint: Let

$$\alpha_n = \frac{b_n - c}{b_n - a_n}, \quad \beta_n = \frac{c - a_n}{b_n - a_n};$$

note that $\alpha_n > 0$, $\beta_n > 0$, $\alpha_n + \beta_n = 1$ and

$$\frac{f(b_n) - f(a_n)}{b_n - a_n} - f'(c) = [g(b_n) - f'(c)]\alpha_n + [g(a_n) - f'(c)]\beta_n,$$

where g is the difference-quotient function of 8.1.1.}

(ii) Show that if $c < a_n < b_n$ the conclusion of (i) may be false. {Hint: Let $b_n = 1/n$, let (a_n) be a sequence with $b_{n+1} < a_n < b_n$ and let $f : [-1, 1] \to \mathbb{R}$ be the piecewise linear function such that $f(1/n) = 1/n^2$, $f(a_n) = 0$ and $f(x) = 0$ for $-1 \le x \le 0$:

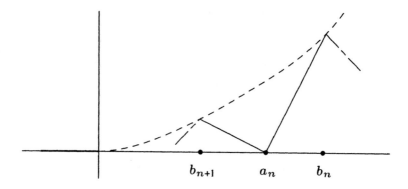

Assume a_n to be chosen so near to b_n that the difference-quotient in (i) is > 1. Note that $f'(0) = 0$.}

8.2. Algebra of Derivatives

The characterization of differentiability in 8.1.5 reduces the proofs of the basic 'laws of differentiation' to elementary algebra:

8.2.1. Theorem. *Let $f : S \to \mathbb{R}$, $g : S \to \mathbb{R}$, where S is a neighborhood of $c \in \mathbb{R}$. If f and g are differentiable at c then so are $f + g$, af ($a \in \mathbb{R}$) and fg, and*

$$(f + g)'(c) = f'(c) + g'(c),$$
$$(af)'(c) = af'(c),$$
$$(fg)'(c) = f(c)g'(c) + f'(c)g(c);$$

if, moreover, $f(c) \neq 0$, *then* $1/f$ *is differentiable at* c *and*

$$(1/f)'(c) = -f'(c)/f(c)^2 \,.$$

Proof. By 8.1.5, there exist functions $A : S \to \mathbb{R}$, $B : S \to \mathbb{R}$, continuous at c, such that

$$f(x) - f(c) = A(x)(x - c), \quad g(x) - g(c) = B(x)(x - c)$$

for all $x \in S$. The function $A + B : S \to \mathbb{R}$ is also continuous at c (5.3.1) and

$$\begin{aligned}
(f + g)(x) - (f + g)(c) &= [f(x) - f(c)] + [g(x) - g(c)] \\
&= A(x)(x - c) + B(x)(x - c) \\
&= (A + B)(x) \cdot (x - c)
\end{aligned}$$

for all $x \in S$; by 8.1.5, $f + g$ is differentiable at c, with derivative $(A + B)(c) = A(c) + B(c) = f'(c) + g'(c)$. The proof for af ($a \in \mathbb{R}$) is similar. For all $x \in S$,

$$\begin{aligned}
f(x)g(x) - f(c)g(c) &= [f(x) - f(c)]g(x) + f(c)[g(x) - g(c)] \\
&= A(x)(x - c)g(x) + f(c)B(x)(x - c)\,,
\end{aligned}$$

thus

$$(fg)(x) - (fg)(c) = [Ag + f(c)B](x) \cdot (x - c)\,,$$

where $Ag + f(c)B$ is continuous at c (8.1.7, 5.3.1); therefore fg is differentiable at c and

$$\begin{aligned}
(fg)'(c) &= [Ag + f(c)B](c) \\
&= A(c)g(c) + f(c)B(c) = f'(c)g(c) + f(c)g'(c)\,.
\end{aligned}$$

Suppose, in addition, that $f(c) \neq 0$. Since f is continuous at c, $1/f$ is defined on a neighborhood T of c (cf. the proof of 7.5.1); for $x \in T$,

$$\begin{aligned}
1/f(x) - 1/f(c) &= [-1/f(c)f(x)][f(x) - f(c)] \\
&= [-1/f(c)f(x)]A(x)(x - c) \\
&= [-A(x)/f(c)f(x)](x - c)\,,
\end{aligned}$$

and the function $B(x) = -A(x)/f(c)f(x)$ is continuous at c (8.1.7, 5.3.3), so $1/f$ is differentiable at c and

$$(1/f)'(c) = B(c) = -f'(c)/f(c)^2 \,. \ \Diamond$$

8.2.2. Remark. There are analogues of 8.2.1 for one-sided differentiability (cf. 8.1.6 and Exercise 2 of §5.5).

Exercises

1. Write out the details for 8.2.2.

2. (i) Let S be a neighborhood of $c \in \mathbb{R}$ and suppose that $f : S \to \mathbb{R}$ is differentiable at c. Prove (by induction): For every positive integer n, the function $g = f^n$ is differentiable at c and $g'(c) = nf^{n-1}(c)f'(c)$. {Hint: $f^n = f \cdot f^{n-1}$.}

(ii) Deduce that a polynomial function $p : \mathbb{R} \to \mathbb{R}$ is differentiable at every $c \in \mathbb{R}$.

3. Prove that, in the notations of 8.2.1, if f and g are differentiable at c and if $g(c) \neq 0$, then f/g is differentiable at c and

$$(f/g)'(c) = \frac{g(c)f'(c) - f(c)g'(c)}{g(c)^2}.$$

4. From Exercises 2 and 3, deduce the differentiability of a rational function p/q at every point c where $q(c) \neq 0$ (cf. 5.3.4).

5. (i) Let $a_0, a_1, \ldots, a_n \in \mathbb{R}$, let $c \in \mathbb{R}$ and let $p : \mathbb{R} \to \mathbb{R}$ be the polynomial function

$$p(x) = \sum_{k=0}^{n} a_k (x - c)^k.$$

Prove: $a_k = p^{(k)}(c)/k!$, where $p^{(k)}$ is the k'th derivative function of p. {Recursively, $p^{(0)} = p$ and $p^{(k)} = (p^{(k-1)})'$. By convention, $0! = 1$.}

(ii) If $p : \mathbb{R} \to \mathbb{R}$ is a polynomial function of degree n and if $c \in \mathbb{R}$, then

$$p(x) = \sum_{k=0}^{n} \frac{p^{(k)}(c)}{k!} (x - c)^k$$

for all $x \in \mathbb{R}$. {Hint: In view of (i), it suffices to show that every power function $x \mapsto x^m$ is a linear combination of functions $x \mapsto (x - c)^k$; look at the binomial expansion

$$x^m = [c + (x - c)]^m = \sum_{k=0}^{m} \binom{m}{k} c^{m-k} (x - c)^k.\}$$

(iii) If $p : \mathbb{R} \to \mathbb{R}$ is a polynomial function such that $p(c) = p'(c) = p''(c) = \ldots = p^{(m)}(c) = 0$, then $p(x) = (x - c)^{m+1} q(x)$ for a suitable polynomial q. {Hint: Use (ii).}

8.3. Composition (Chain Rule)

8.3.1. Theorem. (Chain Rule) *Let $f : S \to \mathbb{R}$, where S is a neighborhood of $c \in \mathbb{R}$; let $g : T \to \mathbb{R}$, where T is a neighborhood of $f(c)$; and* *suppose that* $f(S) \subset T$, *so that the composite function* $g \circ f : S \to \mathbb{R}$ *is defined*:

$$S \xrightarrow{\ f\ } T \xrightarrow{\ g\ } \mathbb{R}$$

$$\begin{array}{ccc} S & & T \\ \cup\!\!\!\!\cup & & \cup\!\!\!\!\cup \\ c & & f(c) \end{array}$$

If f *is differentiable at* c, *and* g *is differentiable at* $f(c)$, *then* $g \circ f$ *is differentiable at* c *and*

$$(g \circ f)'(c) = g'\big(f(c)\big) \cdot f'(c).$$

Proof. Write $h = g \circ f$. By 8.1.5, there exists a function $A : S \to \mathbb{R}$, continuous at c, such that

(1) $$f(x) - f(c) = A(x)(x - c) \quad \text{for all } x \in S.$$

Similarly, there is a function $B : T \to \mathbb{R}$, continuous at $f(c)$, such that

(2) $$g(y) - g\big(f(c)\big) = B(y)\big(y - f(c)\big) \quad \text{for all } y \in T.$$

If $s \in S$ then $f(x) \in T$; putting $y = f(x)$ in (2), we have

$$g\big(f(x)\big) - g\big(f(c)\big) = B\big(f(x)\big) \cdot \big(f(x) - f(c)\big)$$
$$= B\big(f(x)\big) \cdot A(x)(x - c)$$

by (1), thus

(3) $$h(x) - h(c) = [(B \circ f)A](x) \cdot (x - c) \quad \text{for all } x \in S.$$

Since $(B \circ f)A$ is continuous at c (5.6.4, 5.3.1), it follows that h is differentiable at c and

$$h'(c) = [(B \circ f)A](c) = (B \circ f)(c) \cdot A(c)$$
$$= B\big(f(c)\big) \cdot A(c) = g'\big(f(c)\big) \cdot f'(c). \ \Diamond$$

8.3.2. Remarks. There are partial 'one-sided' versions of 8.3.1 (but see Exercise 4). For example, assume S is a right neighborhood of c and T is a neighborhood of $f(c)$; if f is right differentiable at c, and g is differentiable at $f(c)$, then $g \circ f$ is right differentiable at c and

$$(g \circ f)'_r(c) = g'\big(f(c)\big) \cdot f'_r(c).$$

{With notations as in the proof of 8.3.1, A and f are right continuous at c, so $(B \circ f)A$ is right continuous at c (cf. §5.6, Exercise 2 and §5.5, Exercise 2).}

Exercises

1. Write out in full the proofs sketched in 8.3.2.

2. Deduce from 8.3.2 that if $f : S \to \mathbb{R}$ is right differentiable at c and if $f(c) \neq 0$, then $1/f$ is defined in a right neighborhood of c and is right differentiable at c, with

$$(1/f)'_r(c) = -f'_r(c)/f(c)^2 .$$

3. Let S be a right neighborhood of $c \in \mathbb{R}$ and let $T = -S = \{-x : x \in S\}$, which is a left neighborhood of $-c$. If $f : S \to \mathbb{R}$ is right differentiable at c then the function $g : T \to \mathbb{R}$ defined by $g(x) = f(-x)$ is left differentiable at $-c$, and

$$g'_l(-c) = -f'_r(c) .$$

*4. If $f : \mathbb{R} \to \mathbb{R}$ is the function $f(x) = -x$ and $g : \mathbb{R} \to \mathbb{R}$ is the function defined by

$$g(x) = \begin{cases} 0 & \text{for } x \geq 0, \\ x \sin(1/x) & \text{for } x < 0, \end{cases}$$

then g is right differentiable at 0 but $g \circ f$ is not.*

8.4. Local Max and Min

8.4.1. *Definition.* Let $f : S \to \mathbb{R}$, where S is a neighborhood of $c \in \mathbb{R}$. We say that

f has a **local maximum at** c if there exists a neighborhood V of c, with $V \subset S$, such that $f(x) \leq f(c)$ for all $x \in V$;

f has a **local minimum at** c if there exists a neighborhood V of c, with $V \subset S$, such that $f(x) \geq f(c)$ for all $x \in V$ (in other words, $-f$ has a local maximum at c).

8.4.2. *Remarks.* A function $f : S \to \mathbb{R}$ is said to have a **maximum** (or 'global maximum') at $c \in S$ if $f(x) \leq f(c)$ for all $x \in S$ (here S need not be a neighborhood of c); if $f(x) \geq f(c)$ for all $x \in S$ then f is said to have a **minimum** (or 'global minimum') at c. For example, every continuous function defined on a closed interval has a maximum and a minimum (6.3.2).

8.4.3. Lemma. *Suppose* $f : S \to \mathbb{R}$ *has a local maximum at* c. *Then*
(i) f *right differentiable at* c $\Rightarrow$ $f'_r(c) \le 0$;
(ii) f *left differentiable at* c $\Rightarrow$ $f'_l(c) \ge 0$.

Proof. It is implicit that S is a neighborhood of c in $\mathbb{R}$ (8.4.1). Shrinking S if necessary, we can suppose that $f(x) \le f(c)$ for all $x \in S$.
   (i) Let (x_n) be a sequence in S with $x_n > c$ and $x_n \to c$. By assumption,
$$\frac{f(x_n) - f(c)}{x_n - c} \to f'_r(c) .$$
In the difference quotient on the left, $f(x_n) - f(c) \le 0$ and $x_n - c > 0$, so the fraction is ≤ 0, therefore so is its limit (3.4.8).
   (ii) Assuming $x_n < c$, the numerator in the above difference quotient is ≤ 0 and the denominator is < 0. $\Diamond$

8.4.4. Theorem. *Let* $f : S \to \mathbb{R}$, *where* S *is a neighborhood of* $c \in \mathbb{R}$. *If* f *has a local maximum or a local minimum at* c, *and if* f *is differentiable at* c, *then* $f'(c) = 0$.

Proof. If f has a local maximum at c then, by the lemma,
$$0 \le f'_l(c) = f'(c) = f'_r(c) \le 0 ,$$
so $f'(c) = 0$. If f has a local minimum at c, apply the preceding argument to $-f$. $\Diamond$

Exercises

1. Prove that the function $f : \mathbb{R} \to \mathbb{R}$ defined by $f(x) = x^3 - 3x$ has a local maximum at $x = -1$, a local minimum at $x = 1$, but no maximum or minimum value. In contrast, the function $g : [0, 1] \to \mathbb{R}$ defined by $g(x) = x$ has a maximum at 1, a minimum at 0, but no local maximum or local minimum. {Shortcut: After establishing the character of f at 1, note that f is an *odd* function.}

2. True or false (explain): If $f : S \to \mathbb{R}$ has a maximum at an interior point c of S then f has a local maximum at c.

8.5. Mean Value Theorem

   The basic theme of this section is the interaction between a function and its derivative.

   8.5.1. Theorem. (Rolle's theorem) *If* $f : [a, b] \to \mathbb{R}$ *is continuous,* $a < b$, f *is differentiable at every point of* (a, b), *and* $f(a) = f(b)$, *then there exists a point* $c \in (a, b)$ *such that* $f'(c) = 0$.

Proof. By 6.3.1, the range of f is a closed interval, say $f([a,b]) = [m, M]$. If $m = M$ then f is constant and $f'(c) = 0$ for all $c \in (a, b)$. Suppose $m < M$; say $m = f(c)$, $M = f(d)$. Since $f(a) = f(b)$ and $f(c) \neq f(d)$, not both of c and d can be endpoints of $[a, b]$, so at least one of them must be an internal point; if, for example, $d \in (a, b)$, then f has a local maximum at d, so $f'(d) = 0$ by 8.4.4. $\Diamond$

8.5.2. Theorem. (Mean Value Theorem) *If $f : [a, b] \to \mathbb{R}$ is continuous, $a < b$, and f is differentiable at every point of (a, b), then there exists a point $c \in (a, b)$ such that*

$$f(b) - f(a) = f'(c)(b - a).$$

Proof. In other words, there is an internal point at which the tangent line is parallel to the chord joining the endpoints (Figure 11).

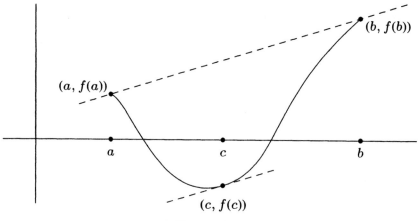

FIGURE 11

Write

$$m = \frac{f(b) - f(a)}{b - a}$$

for the slope of the line joining the endpoints of the graph; the equation of this line is $y = f(a) + m(x - a)$. Define $F : [a, b] \to \mathbb{R}$ by 'subtracting' the line from the graph of f, that is,

$$F(x) = f(x) - [f(a) + m(x - a)].$$

Then F is continuous on $[a, b]$, differentiable on (a, b), $F(a) = 0$, and $F(b) = 0$ (by the definition of m); thus F satisfies the hypotheses of Rolle's theorem, so there is a point $c \in (a, b)$ such that

$$0 = F'(c) = f'(c) - m. \Diamond$$

8.5.3. Corollary. *For a continuous function* $f : [a, b] \to \mathbb{R}$ *that is differentiable on* (a, b), *the following conditions are equivalent:*
(a) f *is increasing*;
(b) $f'(x) \geq 0$ *for all* $x \in (a, b)$.

Proof. (a) $\Rightarrow$ (b): The elementary argument is similar to that in 8.4.3.
(b) $\Rightarrow$ (a): Note first that $f(a) \leq f(b)$; for, in the notation of the mean value theorem, $f'(c) \geq 0$ and $b - a > 0$, therefore $f(b) - f(a) \geq 0$. More generally, if $a \leq x < y \leq b$ then $f(x) \leq f(y)$ (apply the preceding result to the restriction of f to the interval $[x, y]$); in other words, f is increasing. $\Diamond$

8.5.4. Corollary. *If* $f : [a, b] \to \mathbb{R}$ *is continuous, differentiable on* (a, b), *and* $f'(x) = 0$ *for all* $x \in (a, b)$, *then* f *is a constant function.*

Proof. By the preceding corollary, f is increasing; but the hypotheses are also satisfied by $-f$, so f is also decreasing. There's only one way out: f is constant. $\Diamond$

This corollary is greatly strengthened by (ii) of the following theorem (destined for application in the theory of integration):[1]

8.5.5. Theorem. *Let* $f : [a, b] \to \mathbb{R}$ *be a continuous function that is right differentiable at every point of* (a, b).
(i) *If* $f'_r(x) \geq 0$ *for all* $x \in (a, b)$ *then* f *is an increasing function on* $[a, b]$.
(ii) *If* $f'_r(x) = 0$ *for all* $x \in (a, b)$ *then* f *is a constant function.*
(iii) *If* $f'_r(x) > 0$ *for all* $x \in (a, b)$ *then* f *is strictly increasing on* $[a, b]$.

Proof.[2] (iii) We show first that f is increasing. To this end, it suffices to show that $f(a) \leq f(b)$. {For, if $a \leq c < d \leq b$ then the restriction of f to $[c, d]$ also satisfies the hypotheses of (iii).} Assume to the contrary that $f(a) > f(b)$. Choose k so that $f(b) < k < f(a)$ and let

$$A = \{x \in [a, b] : f(x) \geq k\};$$

A is bounded, and nonempty ($a \in A$). Also, A is a closed set; for, if $x_n \in A$ and $x_n \to x$, then $x \in [a, b]$ (because $[a, b]$ is a closed set) and $f(x) = \lim f(x_n) \geq k$ by the continuity of f and (8) of 3.4.8, thus $x \in A$. It follows that A has a largest element c, namely $c = \sup A$ (4.5.7). In particular, $f(c) \geq k$ and, since $f(b) < k$, we have $c \neq b$, therefore $c < b$. We are going to show that also $f(c) \leq k$, hence $f(c) = k$.

---

[1]The rest of the section, somewhat difficult, can be deferred until 9.6.11. It would not be a sin to omit it altogether.
[2]A combination of ideas I learned from my colleague Ralph Showalter and the book of E. J. McShane [*Integration*, Princeton, 1944], p. 200, 34.1.

Choose a sequence (t_n), $c < t_n < b$, such that $t_n \to c$. Since c is the largest element of A and $t_n > c$, t_n does not belong to A, therefore $f(t_n) < k$; passing to the limit, $f(c) \le k$. Thus $f(c) = k$ and $f(b) < f(c) < f(a)$; in particular, $c \ne a$, therefore $a < c < b$, that is, $c \in (a, b)$. By hypothesis, f is right differentiable at c and $f'_r(c) > 0$. However, $f(t_n) - f(c) = f(t_n) - k < 0$, and passage to the limit in the inequality

$$\frac{f(t_n) - f(c)}{t_n - c} < 0$$

yields $f'_r(c) \le 0$, a contradiction.

We now know that f is increasing. If it failed to be strictly increasing, there would exist a pair of points c, d in $[a, b]$ with $c < d$ and $f(c) = f(d)$; it would then follow that f is constant on $[c, d]$, so that $f'_r(x) = 0$ for all $x \in (c, d)$, contrary to hypothesis.

(i) Assuming $a \le c < d \le b$, we are to show that $f(c) \le f(d)$. Given any positive integer n, it suffices to show that $f(c) + c/n < f(d) + d/n$ (passage to the limit then yields the desired inequality). Let $g : [a, b] \to \mathbb{R}$ be the function defined by $g(x) = f(x) + x/n$. Then g is continuous and, for every $x \in (a, b)$, g is right differentiable at x with $g'_r(x) = f'_r(x) + 1/n \ge 1/n > 0$; by (iii), $g(c) < g(d)$, that is, $f(c) + c/n < f(d) + d/n$.

(ii) By assumption, $f'_r(x) = 0$ and $(-f)'_r(x) = -f'_r(x) = 0$ for all $x \in (a, b)$. By (i), f and $-f$ are both increasing, so f is both increasing and decreasing—in other words, constant. $\Diamond$

8.5.6. *Remark.* The foregoing proof of (iii) shows that, to conclude f is strictly increasing, it suffices to assume that $f'_r(x) \ge 0$ for all $x \in (a, b)$ and that every open subinterval (c, d) of $[a, b]$ contains a point x with $f'_r(x) > 0$.

8.5.7. *Remark.* Theorem 8.5.5 is true with "right" replaced by "left" and f'_r by f'_l. {Apply 8.5.5 to the function $x \mapsto -f(-x)$ on $[-b, -a]$ (or to the function $x \mapsto -f(a + b - x)$ on $[a, b]$).}

Exercises

1. Write out in full the proof sketched in 8.5.7. {Hint: Cf. §8.3, Exercise 3.}

2. Let I be an interval that is an open subset of $\mathbb{R}$ and let $f : I \to \mathbb{R}$ be a function such that, at every point $x \in I$, f has a derivative $f'(x) \ne 0$. Let $J = f(I)$ (also an interval, by 6.1.2). Prove:

(i) f is injective (hence strictly monotone). {Hint: Rolle's theorem (and §6.4, Exercise 7).}

(ii) The interval J is also an open subset of $\mathbb{R}$.

(iii) The function $g : J \to I$ inverse to f is differentiable at every point $y \in J$, and $g'(y) = 1/f'(g(y))$. {Hint: g is continuous (6.5.2).}

3. Suppose $f : I \to \mathbb{R}$ is continuous, I an interval (of any sort), and suppose that at every interior point x of I, f is differentiable and $f'(x) \geq 0$. Prove: f is an increasing function on I.

4. Let $a < c < b$. Suppose $f : (a, b) \to \mathbb{R}$ is continuous, differentiable at every point of $(a, b) - \{c\}$, and that f' has a limit at c. Prove: f is differentiable at c and

$$f'(c) = \lim_{x \to c} f'(x)$$

(thus f' is continuous at c). {Hint: MVT.}

5. Suppose $f : [a, b] \to \mathbb{R}$, $g : [a, b] \to \mathbb{R}$ are continuous, f and g are differentiable on (a, b), and $g'(x) \neq 0$ for all $x \in (a, b)$. Prove there exists a point $c \in (a, b)$ such that

$$\frac{f(b) - f(a)}{g(b) - g(a)} = \frac{f'(c)}{g'(c)}$$

(*Cauchy's mean value theorem*).

{Hint: The fraction on the left exists by Rolle's theorem. Speaking of Rolle's theorem, consider the function $F = [f(b) - f(a)]g - [g(b) - g(a)]f$.}

6. (*Darboux's theorem*) If the interval I is an open subset of $\mathbb{R}$, and if $f : I \to \mathbb{R}$ is differentiable at every point of I, then the range of f' is an interval (not necessarily an open set). (This has the flavor of an 'intermediate value theorem' for f', but we are not assuming that f' is continuous!)

{Hint: If a, b are points of I with $a < b$, and if K lies strictly between $f'(a)$ and $f'(b)$, argue that the function $g : [a, b] \to \mathbb{R}$ defined by $g(x) = f(x) - Kx$ has a maximum or a minimum in (a, b); if, for example, $f'(a) < K < f'(b)$, inspection of the signs of the one-sided derivatives of g at a and b shows that g has a minimum at an internal point of $[a, b]$; cf. 8.4.4.}

7. (*L'Hospital's rule*) If $f(x) \to 0$, $g(x) \to 0$ and $f'(x)/g'(x) \to L$ as $x \to c$, then $f(x)/g(x) \to L$ as $x \to c$. The hypotheses on f and g: on some deleted neighborhood of c, f and g are differentiable (hence continuous) and g' is zero-free, and the stated limits of f, g and f'/g' exist. The conclusion is that g is zero-free on a deleted neighborhood of c and that f/g has a limit, as $x \to c$, equal to the limit of f'/g'.

{Hint: Since the hypotheses and conclusion are 'local', we can suppose that f and g are defined on an open interval (a, b) with $a < c < b$, that f', g' are defined on $(a, b) - \{c\}$ and g' is never 0 there, and that $f(c) = g(c) = 0$ (if necessary, define—or redefine— f and g at c to have this property), so that f and g are continuous on all of (a, b).

Note that g has no zero in (a, b) other than c (Rolle). If $x_n \to c$, $x_n \neq c$, Exercise 5 provides a point t_n between c and x_n such that $f(x_n)/g(x_n) = f'(t_n)/g'(t_n)$; necessarily $t_n \to c$.}

8. Let $f : [1, 3] \to \mathbb{R}$ be a continuous function that is differentiable on $(1, 3)$ with derivative $f'(x) = [f(x)]^2 + 4$ for all $x \in (1, 3)$. True or false (explain): $f(3) - f(1) = 5$.

9. The argument of 8.5.5, (iii) can be sharpened in the following way. Assume that $f : [a, b] \to \mathbb{R}$ is continuous and that N is a subset of $[a, b]$ such that, for every $x \in (a, b) - N$, f is right differentiable at x and $f'_r(x) > 0$.

(iii') If $f(N)$ has empty interior, then f is increasing; if both N and $f(N)$ have empty interior, then f is strictly increasing.

{Hint: In the argument of 8.5.5, if $f(a) > f(b)$, then $f(N)$ cannot contain the interval $\big(f(b), f(a)\big)$; choose $k \in \big(f(b), f(a)\big) - f(N)$.}

10. Let $f : [a, b] \to \mathbb{R}$ be continuous and suppose (c_n) is a sequence of points of $[a, b]$ such that f is right differentiable at every point of (a, b) not equal to any of the points c_n. Let $N = \{c_n : n \in \mathbb{P}\}$. Then the statements (i)–(iii) of 8.5.5 remain true with $x \in (a, b)$ replaced by $x \in (a, b) - N$.

{Hint: The sets N and $f(N)$ both have empty interior (cf. §2.6, Exercise 4).}

11. Corollary 8.5.3 can be generalized as follows. Let $f : [a, b] \to \mathbb{R}$ be continuous and suppose there exists a sequence (c_n) of points of $[a, b]$ such that f is right differentiable at every point of (a, b) other than the c_n. The following conditions are equivalent:

(a) f is increasing on $[a, b]$;
(b) $f'_r(x) \geq 0$ for all $x \in (a, b) - \{c_n : n \in \mathbb{P}\}$.
{Hint: Exercise 10.}

Riemann Integral

To save repetition, the following notations are fixed for the entire chapter:

$$[a,b] \text{ is a closed interval of } \mathbb{R}, \ a < b$$
$$f : [a,b] \to \mathbb{R} \text{ is a } \textbf{bounded} \text{ function}$$
$$M = \sup f = \sup\{f(x) : \ a \le x \le b\}$$
$$m = \inf f = \inf\{f(x) : \ a \le x \le b\}.$$

We also write $M = M(f)$ and $m = m(f)$ to indicate the dependence of M and m on f. Other notations are introduced as needed (for subintervals of $[a,b]$, for other functions, etc.), but the foregoing are the indispensable core.

The agenda: When should f have an 'integral' and what should it be? What are its properties? What good is it?

9.1. Upper and Lower Integrals: The Machinery

9.1.1. *Definition.* A **subdivision**[1] σ of $[a,b]$ is a finite list of points, starting at a, increasing strictly, and ending at b:

$$\sigma = \{a = a_0 < a_1 < a_2 < \cdots < a_n = b\}.$$

---

[1]The term 'partition' is frequently used. However, the most firmly established use of the word 'partition' is for the decomposition of a set into pairwise disjoint subsets, whereas adjacent subintervals associated with a subdivision have a point in common; hence our preference for the term 'subdivision'.

The a_ν $(\nu = 0, 1, 2, \ldots, n)$ are called the *points* of the subdivision; the 'trivial subdivision' $\sigma = \{a = a_0 < a_1 = b\}$ is allowed. The effect of σ (when $n > 1$) is to break up the interval $[a, b]$ into n *subintervals*

$$[a_0, a_1], [a_1, a_2], \ldots, [a_{n-1}, a_n];$$

the length of the ν'th subinterval is denoted e_ν,

$$e_\nu = a_\nu - a_{\nu-1} \quad (\nu = 1, \ldots, n);$$

the largest of these lengths is called the **norm** of the subdivision σ, written

$$N(\sigma) = \max\{e_\nu : \nu = 1, \ldots, n\}.$$

9.1.2. Definition. With the preceding notations, for $\nu = 1, \ldots, n$ we write
$$M_\nu = \sup\{f(x) : a_{\nu-1} \le x \le a_\nu\},$$
$$m_\nu = \inf\{f(x) : a_{\nu-1} \le x \le a_\nu\};$$

thus M_ν and m_ν are the supremum and infimum of the function $f|[a_{\nu-1}, a_\nu]$ (the restriction of f to the ν'th subinterval). Obviously $m_\nu \le M_\nu$; the difference

$$\omega_\nu = M_\nu - m_\nu \quad (\ge 0)$$

is called the **oscillation** of f over the subinterval $[a_{\nu-1}, a_\nu]$.[2] When we wish to show the dependence of these numbers on f, we write $M_\nu(f)$, $m_\nu(f)$, $\omega_\nu(f)$.

9.1.3. Definition. With the preceding notations, the **upper sum** of f for the subdivision σ is the number

$$S(\sigma) = \sum_{\nu=1}^{n} M_\nu e_\nu$$

and the **lower sum** of f for σ is the number

$$s(\sigma) = \sum_{\nu=1}^{n} m_\nu e_\nu;$$

we write $S_f(\sigma)$ and $s_f(\sigma)$ to express the dependence of these numbers on f and σ.

---

[2]This concept is not needed until §9.7.

The upper and lower sums can (at least for $f \geq 0$) be interpreted as crude 'rectangular' approximations to the area[3] under the graph of f. For example, in Figure 12,

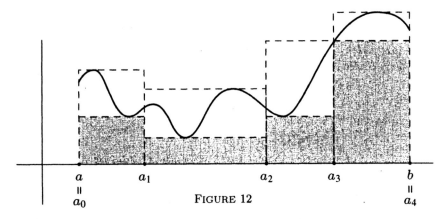

FIGURE 12

the shorter rectangles (shaded) represent the terms of the lower sum

$$m_1 e_1 + m_2 e_2 + m_3 e_3 + m_4 e_4$$

and the taller rectangles, the upper sum

$$M_1 e_1 + M_2 e_2 + M_3 e_3 + M_4 e_4$$

for a subdivision $\sigma = \{a = a_0 < a_1 < a_2 < a_3 < a_4 = b\}$.

9.1.4. Theorem. *If σ is any subdivision of $[a, b]$, then*

$$m(b - a) \leq s(\sigma) \leq S(\sigma) \leq M(b - a).$$

Proof. Say $\sigma = \{a = a_0 < a_1 < \cdots < a_n = b\}$. For $\nu = 1, \ldots, n$,

$$(*) \qquad\qquad m \leq m_\nu \leq M_\nu \leq M;$$

for, $m_\nu \leq M_\nu$ is obvious, and the inequalities $m \leq m_\nu$, $M_\nu \leq M$ follow from the inclusion

$$\{f(x) : a_{\nu-1} \leq x \leq a_\nu\} \subset \{f(x) : a \leq x \leq b\}.$$

Multiply (*) by e_ν and sum over ν, noting that

$$\sum_{\nu=1}^{n} e_\nu = b - a. \ \diamondsuit$$

---

[3]The geometric language is pure fantasy, albeit helpful; what is actually happening takes place in the definitions and theorems. One consequence of the theory is a definition of 'area' in which we can have some confidence.

It follows that the sets

$$\{s(\sigma) : \ \sigma \text{ any subdivision of } [a, b]\},$$
$$\{S(\sigma) : \ \sigma \text{ any subdivision of } [a, b]\},$$

are bounded; indeed, by 9.1.4 they are subsets of the interval $[m(b - a), M(b - a)]$. If, in a sense, the 'right answer' is the area under the graph of f, then the lower sums are too small and the upper sums are too big; the appropriate numbers to consider are the following:

9.1.5. Definition. The **lower integral** of f over $[a, b]$ is defined to be the supremum of the lower sums, written

$$\underline{\int_a^b} f = \sup\{s(\sigma) : \ \sigma \text{ any subdivision of } [a, b]\},$$

and the **upper integral** is defined to be the infimum of all the upper sums, written

$$\overline{\int_a^b} f = \inf\{S(\sigma) : \ \sigma \text{ any subdivision of } [a, b]\}.$$

9.1.6. Example. In favorable cases these two numbers are equal, but here's a ghastly example where everything goes wrong: $f(x) = 1$ for all rational x in $[a, b]$ and $f(x) = 0$ for all irrational x. For this function, every lower sum is 0 and every upper sum is $b - a$, thus

$$\underline{\int_a^b} f = 0 \quad \text{and} \quad \overline{\int_a^b} f = b - a.$$

9.1.7. Remarks. There's a useful analogy with bounded sequences (§3.7). In arriving at the upper integral, for each subdivision σ we take a supremum (actually, one for each term of $S(\sigma)$), then we take the infimum of the $S(\sigma)$ over all possible subdivisions σ; this is analogous to the *limit superior* of a bounded sequence (3.7.2). Similarly, the definition of lower integral is analogous to the limit inferior of a bounded sequence (inf followed by sup). The pathology in 9.1.6 represents a sort of 'divergence'; just as the 'nice' bounded sequences are the convergent ones (those for which $\liminf = \limsup$), the 'nice' bounded functions should, by analogy, be those for which the lower integral is equal to the upper integral. (All of these prophecies are fulfilled in what follows!)

Obviously

$$s(\sigma) \le \underline{\int_a^b} f \quad \text{and} \quad \overline{\int_a^b} f \le S(\sigma)$$

for every subdivision σ. Some other useful inequalities:

9.1.8. **Theorem.** *For every bounded function* $f : [a, b] \to \mathbb{R}$,

$$m(b - a) \leq \int_{\underline{\quad} a}^{b} f \leq M(b - a),$$

$$m(b - a) \leq \int_{a}^{\overline{\quad} b} f \leq M(b - a),$$

where $m = \inf f$ *and* $M = \sup f$.

Proof. This is obvious from 9.1.4. ◊

Upper and lower sums are in a sense approximations to the upper and lower integrals. The way to improve the approximation is to make the subdivision 'finer' in the following sense:

9.1.9. *Definition.* Let σ and τ be subdivisions of $[a, b]$. We say that τ **refines** σ (or that τ is a *refinement* of σ)—written $\tau \succ \sigma$ (or $\sigma \prec \tau$)—if every point of σ is also a point of τ. Thus, if

$$\sigma = \{a = a_0 < a_1 < \cdots < a_n = b\},$$
$$\tau = \{a = b_0 < b_1 < \cdots < b_m = b\},$$

then $\tau \succ \sigma$ means that each a_ν is equal to some b_μ; in other words, as sets,

$$\{a_0, a_1, \ldots, a_n\} \subset \{b_0, b_1, \ldots, b_m\}.$$

9.1.10. *Remarks.* Trivially $\sigma \succ \sigma$; if $\rho \succ \tau$ and $\tau \succ \sigma$ then $\rho \succ \sigma$. If $\tau \succ \sigma$ and $\sigma \succ \tau$—that is, σ and τ have the same points—then σ and τ are the same subdivision and we write $\sigma = \tau$.

If $\tau \succ \sigma$ it is obvious that $N(\tau) \leq N(\sigma)$. The effect of refinement on upper and lower sums is as follows:

9.1.11. **Lemma.** *If* $\tau \succ \sigma$ *then* $S(\tau) \leq S(\sigma)$ *and* $s(\tau) \geq s(\sigma)$.

Proof. In other words, refinement can only shrink (or leave fixed) an upper sum; refinement can only increase (or leave fixed) a lower sum.

If $\tau = \sigma$ there is nothing to prove. Otherwise, if τ has $r \geq 1$ points not in σ, we can start at σ and arrive at τ in r steps by inserting one of these points at a time, say

$$\sigma = \sigma_0 \prec \sigma_1 \prec \ldots \prec \sigma_r = \tau,$$

where σ_k is obtained from σ_{k-1} by inserting one new point. We need only show that $S(\sigma_k) \leq S(\sigma_{k-1})$ and $s(\sigma_k) \geq s(\sigma_{k-1})$; thus, it suffices to consider the case that τ is obtained from σ by adding one new point c. Suppose

$$\sigma = \{a = a_0 < a_1 < \cdots < a_n = b\}.$$

Say c belongs to the μ'th subinterval, $a_{\mu-1} < c < a_\mu$, so that

$$\tau = \{a = a_0 < a_1 < \ldots < a_{\mu-1} < c < a_\mu < a_{\mu+1} < \cdots < a_n = b\}.$$

The terms of $S(\tau)$ are the same as those of $S(\sigma)$ except that the μ'th term of $S(\sigma)$ is replaced by two terms of $S(\tau)$; thus, in calculating $S(\sigma) - S(\tau)$ (we are trying to show that it is ≥ 0) all of the action is in the μ'th term of $S(\sigma)$, the other terms of $S(\sigma)$ being canceled out by terms of $S(\tau)$. In other words (replacing f by its restriction to $[a_{\mu-1}, a_\mu]$), we are reduced to the case that

$$\sigma = \{a < b\}, \quad \tau = \{a < c < b\}.$$

Writing $M = \sup f$ as before, and

$$M' = \sup\{f(x): \ a \leq x \leq c\},$$
$$M'' = \sup\{f(x): \ c \leq x \leq b\},$$

we have

$$S(\sigma) = M(b-a) \quad \text{and} \quad S(\tau) = M'(c-a) + M''(b-c);$$

obviously $M' \leq M$ and $M'' \leq M$, so

$$S(\tau) \leq M(c-a) + M(b-c) = M(b-a) = S(\sigma),$$

thus $S(\tau) \leq S(\sigma)$. Similarly $s(\tau) \geq s(\sigma)$ (with the obvious notations, $m \leq m'$ and $m \leq m''$, etc.). $\Diamond$

The middle inequality in 9.1.4 can be improved in an interesting way (cf. the 'alternative proof' of 3.7.4):

9.1.12. Lemma. *If* σ *and* τ *are any two subdivisions of* $[a, b]$, *then* $s(\sigma) \leq S(\tau)$.

Proof. Let ρ be a subdivision such that $\rho \succ \sigma$ and $\rho \succ \tau$. (Such a ρ is called a *common refinement* of σ and τ; for example, create ρ by lumping together all of the points of σ and τ.) By 9.1.11 and 9.1.4,

$$s(\sigma) \leq s(\rho) \leq S(\rho) \leq S(\tau). \ \Diamond$$

The analogue of 3.7.4 (lim inf $\leq$ lim sup):

9.1.13. Theorem. *For every bounded function* $f: [a, b] \to \mathbb{R}$,

$$\int_{\underline{a}}^b f \leq \int_a^{\overline{b}} f.$$

Proof. Fix a subdivision τ. By 9.1.12, $s(\sigma) \leq S(\tau)$ for every subdivision σ, therefore

$$(*) \qquad \int_{\underline{a}}^{b} f \leq S(\tau)$$

by the definition of lower integral (as the *least* upper bound of the set of all lower sums). Now let τ vary: the validity of $(*)$ for every τ implies that

$$\int_{\underline{a}}^{b} f \leq \overline{\int_{a}^{b}} f$$

by the definition of the upper integral (as the *greatest* lower bound of the set of all upper sums). $\diamond$

Exercises

1. If $f \geq 0$ then all upper and lower sums for f are ≥ 0, and

$$0 \leq \int_{\underline{a}}^{b} f \leq \overline{\int_{a}^{b}} f .$$

2. If $f : [a, b] \to \mathbb{R}$ and $g : [a, b] \to \mathbb{R}$ are bounded functions such that $f(x) \leq g(x)$ for all $x \in [a, b]$, then

$$\int_{\underline{a}}^{b} f \leq \int_{\underline{a}}^{b} g \quad \text{and} \quad \overline{\int_{a}^{b}} f \leq \overline{\int_{a}^{b}} g .$$

{Hint: Compare $s_f(\sigma)$ and $s_g(\sigma)$ for any subdivision σ.}

3. Suppose $a < c < b$, $f(c) = 1$, and $f(x) = 0$ for all $x \in [a, b] - \{c\}$. Compute the lower and upper integrals of f. Do the same for the function f such that $f(x) = 0$ for $x \in [a, c)$ and $f(x) = 1$ for $x \in [c, b]$.

4. Prove that if $f : [a, b] \to \mathbb{R}$ is increasing then its lower and upper integrals are equal. (Similarly if f is decreasing.)

{Hint: For any subdivision σ (with notations as in 9.1.3), calculate m_ν, M_ν and $S(\sigma) - s(\sigma)$. Assuming the points of subdivision are *equally spaced*, look at the chain of inequalities

$$s(\sigma) \leq \int_{\underline{a}}^{b} f \leq \overline{\int_{a}^{b}} f \leq S(\sigma)$$

and calculate the gap between the left and right ends.}

5. Fix an increasing function $g : [a, b] \to \mathbb{R}$. As usual, $f : [a, b] \to \mathbb{R}$ is any bounded function. In the notations of 9.1.1 and 9.1.2, define $N(\sigma)$ as before (as the maximal length of a subinterval of σ), then make one change: define $e_\nu = g(a_\nu) - g(a_{\nu-1})$. Since g is increasing, $e_\nu \geq 0$; also,

$$\sum_{\nu=1}^{n} e_\nu = g(b) - g(a), \quad s_f(\sigma) = \sum_{\nu=1}^{n} m_\nu [g(a_\nu) - g(a_{\nu-1})], \text{ etc.}$$

The exercise: work through the section taking into account the revised definition of e_ν. In this context, the notations of 9.1.5 are replaced by

$$\int_{\underline{a}}^{b} f \, dg \quad \text{and} \quad \overline{\int_{a}}^{b} f \, dg,$$

called the *Riemann-Stieltjes* lower and upper integrals of f *with respect to* g.[4]

6. If $f : [a, b] \to \mathbb{R}$ is the given bounded function and $g : [-b, -a] \to \mathbb{R}$ is defined by $g(y) = f(-y)$ for $-b \leq y \leq -a$, then

$$\overline{\int_{a}}^{b} f = \overline{\int_{-b}}^{-a} g$$

and similarly for the lower integrals.

{Hint: If $\sigma = \{a = a_0 < a_1 < \cdots < a_n = b\}$, write $-\sigma = \{-b = b_0 < b_1 < \cdots < b_n = -a\}$, where $b_\nu = -a_{n-\nu}$ for $\nu = 0, 1, \ldots, n$. For every ν,

$$\{g(y) : b_{\nu-1} \leq y \leq b_\nu\} = \{f(x) : a_{n-\nu} \leq x \leq a_{n-\nu+1}\},$$

whence $M_\nu(g) = M_{n-\nu+1}(f)$, $S_g(-\sigma) = S_f(\sigma)$, etc.}

9.2. First Properties of Upper and Lower Integrals

The following theorem cuts the labor in half by reducing the study of lower integrals to that of upper integrals:

9.2.1. Theorem. *For every bounded function* $f : [a, b] \to \mathbb{R}$,

$$\int_{\underline{a}}^{b} f = - \overline{\int_{a}}^{b} (-f).$$

---

[4]Cf. the books of W. Rudin [*Principles of mathematical analysis*, p. 120, 3rd. edn., McGraw-Hill, New York, 1976] and A. Devinatz [*Advanced calculus*, p. 210, Holt, Rinehart and Winston, New York, 1968].

Proof. Let σ be any subdivision of $[a, b]$ and adopt the notations of 9.1.3. Writing

$$A_\nu = \{f(x) : \ a_{\nu-1} \leq x \leq a_\nu \},$$

we have

$$\sup(-A_\nu) = -(\inf A_\nu)$$

(1.3.8); this means that, in the notations of 9.1.2,

$$M_\nu(-f) = -m_\nu(f)$$

for $\nu = 1, \ldots, n$, whence

$$S_{-f}(\sigma) = -s_f(\sigma).$$

Writing

$$B = \{s_f(\sigma) : \ \sigma \ \text{any subdivision of} \ [a, b]\},$$

we have

$$-B = \{S_{-f}(\sigma) : \ \sigma \ \text{any subdivision of} \ [a, b]\},$$

therefore

$$\underline{\int}_a^b f = \sup B = -\inf(-B) = -\overline{\int}_a^b (-f). \ \Diamond$$

9.2.2. *Definition.* If $a \leq c < d \leq b$, the definitions for f can be applied to the restriction $f \| [c, d]$ of f to $[c, d]$, that is, to the function $x \mapsto f(x)$ $(c \leq x \leq d)$; instead of the ponderous notations

$$\underline{\int}_c^d f \| [c, d] \quad \text{and} \quad \overline{\int}_c^d f \| [c, d]$$

we write simply

$$\underline{\int}_c^d f \quad \text{and} \quad \overline{\int}_c^d f.$$

It is also convenient to define

$$\underline{\int}_c^c f = \overline{\int}_c^c f = 0$$

for any $c \in [a, b]$ (not unreasonable when one thinks of area!).

With these conventions, the upper and lower integral is (for a fixed function f) an 'additive function of the endpoints of integration':

9.2.3. *Theorem.* *If* $a \leq c \leq b$ *then*

(i)
$$\overline{\int}_a^b f = \overline{\int}_a^c f + \overline{\int}_c^b f,$$

(ii)
$$\int_{\underline{a}}^{b} f = \int_{\underline{a}}^{c} f + \int_{\underline{c}}^{b} f .$$

Proof. Both equations are trivial when $c = a$ or $c = b$; we can suppose $a < c < b$. In view of 9.2.1, it suffices to prove (i); writing L for the left side of (i) and R for the right side, let us show that $L \leq R$ and $L \geq R$.

Proof that $L \leq R$: Let σ_1 be any subdivision of $[a, c]$, σ_2 any subdivision of $[c, b]$, and write $\sigma = \sigma_1 \oplus \sigma_2$ for the subdivision of $[a, b]$ obtained by joining σ_1 and σ_2 at their common point c. It is then clear that

$$S(\sigma) = S(\sigma_1) + S(\sigma_2)$$

(the upper sum on the left pertains to f, those on the right pertain to the restrictions of f to $[a, c]$ and $[c, b]$); then

$$\int_{a}^{\overline{b}} f \leq S(\sigma) = S(\sigma_1) + S(\sigma_2) ,$$

so

$$\int_{a}^{\overline{b}} f - S(\sigma_1) \leq S(\sigma_2) .$$

Varying σ_2 over all possible subdivisions of $[c, b]$, it follows that

$$\int_{a}^{\overline{b}} f - S(\sigma_1) \leq \int_{c}^{\overline{b}} f ,$$

thus

$$\int_{a}^{\overline{b}} f - \int_{c}^{\overline{b}} f \leq S(\sigma_1) ;$$

since this is true for all σ_1 we have

$$\int_{a}^{\overline{b}} f - \int_{c}^{\overline{b}} f \leq \int_{a}^{\overline{c}} f ,$$

whence $L \leq R$.

Proof that $L \geq R$: Let σ be any subdivision of $[a, b]$. Let τ be a subdivision of $[a, b]$ such that $\tau \succ \sigma$ and τ includes the point c (for example, let τ be the result of inserting c into σ if it's not already there). Since c is a point of τ, as in the first part of the proof we can write $\tau = \tau_1 \oplus \tau_2$ with τ_1 a subdivision of $[a, c]$ and τ_2 a subdivision of $[c, b]$. Then

$$S(\sigma) \geq S(\tau) = S(\tau_1) + S(\tau_2) \geq \int_{a}^{\overline{c}} f + \int_{c}^{\overline{b}} f$$

(the first inequality follows from $\tau \succ \sigma$, the second from the definition of upper integral). Thus $S(\sigma) \geq R$ for every subdivision σ of $[a,b]$, whence $L \geq R$. $\Diamond$

Exercises

1. (i) If $c > 0$ then

$$\int_a^{\overline{b}} cf = c \int_a^{\overline{b}} f$$

and similarly for the lower integral.

(ii) If $c < 0$ then

$$\int_a^{\overline{b}} cf = c \int_{\underline{a}}^b f .$$

{Hint: $cf = -(-c)f$.}

2. If $a < c < b$ and

$$f(x) = \begin{cases} 0 & \text{for } a \leq x < c \\ 1 & \text{for } c \leq x \leq b \end{cases}$$

then the upper and lower integrals of f are equal (§9.1, Exercise 3). Derive their value using 9.2.3.

3. The upper and lower integrals of the function $f(x) = x$ are equal (§9.1, Exercise 4); find their value.

{Hint: Let σ_n be the subdivision of $[a,b]$ into n subintervals of equal length $h = (b-a)/n$. Write out the expression for the lower sum $s(\sigma_n)$, simplify it using the formula $1 + 2 + 3 + \ldots + (n-1) = \frac{1}{2}(n-1)n$, and show that $s(\sigma_n) \uparrow \frac{1}{2}(b^2 - a^2)$; infer that the lower integral is $\geq \frac{1}{2}(b^2 - a^2)$. Then look at the upper sums $S(\sigma_n)$.}

4. The results of this section extend to Riemann-Stieltjes upper and lower integrals. {Caution: Exercise 2 does not carry over without an added assumption on the function g of §9.1, Exercise 5.}

9.3. Indefinite Upper and Lower Integrals

The notations of 9.2.2 allow us to regard the upper and lower integrals as functions of the endpoints of integration.

9.3.1. Definition. For the given bounded function $f : [a,b] \to \mathbb{R}$, we define functions $F : [a,b] \to \mathbb{R}$ and $H : [a,b] \to \mathbb{R}$ by the formulas

$$F(x) = \int_a^{\overline{}x} f \qquad (a \le x \le b),$$

$$H(x) = \int_{\underline{}a}^x f \qquad (a \le x \le b).$$

(Later in the section we consider variable lower endpoints of integration, leading to a function G complementary to F, and a function K complementary to H.) The function F is called the **indefinite upper integral** of f, and H is called the **indefinite lower integral** of f.

By convention (9.2.2), $F(a) = H(a) = 0$; by 9.1.13, $H(x) \le F(x)$ for all $x \in [a,b]$. The pleasant surprise of this section is that the functions F and H have nice properties even if *nothing* is assumed about the given bounded function f; and (exaggerating a little) every nice property of f (like continuity) yields an even nicer property of F (like differentiability). First, a property that comes free of charge:

9.3.2. Theorem. *Let* $k = \max\{|m|, |M|\}$, *where* $m = \inf f$ *and* $M = \sup f$. *Then*

$$|F(x) - F(y)| \le k|x - y|, \quad |H(x) - H(y)| \le k|x - y|$$

for all x, y *in* $[a, b]$; *in particular*, F *and* H *are continuous on* $[a, b]$.

Proof. We can suppose $x < y$. By the 'additivity' proved in 9.2.3,

$$\int_a^{\overline{}y} f = \int_a^{\overline{}x} f + \int_x^{\overline{}y} f,$$

thus

$$\int_x^{\overline{}y} f = F(y) - F(x).$$

If m' and M' are the infimum and supremum of f on the interval $[x, y]$, we have $m \le m' \le M' \le M$; citing 9.1.8,

$$m(y - x) \le m'(y - x) \le \int_x^{\overline{}y} f \le M'(y - x) \le M(y - x),$$

thus

$$m(y - x) \le F(y) - F(x) \le M(x - y).$$

Since $|m| \le k$ and $|M| \le k$, so that $-k \le m$ and $M \le k$, it follows that

$$-k(y - x) \le F(y) - F(x) \le k(y - x),$$

that is, $|F(y) - F(x)| \le k(y - x) = k|y - x|$.

The proof for H is similar (or take a shortcut via 9.2.1). ◊

9.3.3. Theorem. *If $f \geq 0$ then F and H are increasing functions.*

Proof. It is clear from 9.1.8 that the upper and lower integrals of a nonnegative function are nonnegative. If $a \leq c < d \leq b$ then, citing 9.2.3,

$$F(d) = F(c) + \int_c^d f \geq F(c),$$

thus F is increasing. Similarly for H . $\Diamond$

To hold down the lengths of proofs, it is sometimes convenient to separate out 'right' and 'left' versions:

9.3.4. Theorem. *If $a \leq c < b$ and f is right continuous at c, then F and H are right differentiable at c and $F_r'(c) = H_r'(c) = f(c)$.*

Proof. We give the proof for F; the proof for H is entirely similar. Let $\epsilon > 0$. We seek a $\delta > 0$, with $c + \delta < b$, such that

(∗) $c < x < c + \delta \ \Rightarrow \ \left| \dfrac{F(x) - F(c)}{x - c} - f(c) \right| \leq \epsilon.$

Since f is right continuous at c, there exists a $\delta > 0$, with $c + \delta < b$, such that

$$c \leq t \leq c + \delta \ \Rightarrow \ |f(t) - f(c)| \leq \epsilon.$$

$$x$$

FIGURE 13

Let $c < x < c + \delta$ (see Figure 13). For all $t \in [c, x]$, we have $|f(t) - f(c)| \leq \epsilon$, so

$$f(c) - \epsilon \leq f(t) \leq f(c) + \epsilon.$$

If m_x and M_x are the infimum and supremum of f on $[c, x]$, then

$$f(c) - \epsilon \leq m_x \leq M_x \leq f(c) + \epsilon,$$

therefore

$$[f(c) - \epsilon](x - c) \leq m_x(x - c) \leq \int_c^x f \leq M_x(x - c) \leq [f(c) + \epsilon](x - c).$$

Citing 9.2.3, we thus have

$$[f(c) - \epsilon](x - c) \leq F(x) - F(c) \leq [f(c) + \epsilon](x - c),$$

whence (∗). $\Diamond$

The 'left' version of 9.3.4:

9.3.4.' Theorem. *If* $a < c \le b$ *and* f *is left continuous at* c, *then* F *and* H *are left differentiable at* c *and* $F_l'(c) = H_l'(c) = f(c)$.

Proof. The easiest strategy is to modify the preceding proof: replace $c < x < c + \delta$ by $c - \delta < x < c$, $[c, x]$ by $[x, c]$, etc.

{An alternative strategy is to apply 9.3.4 to the function $g : [-b, -a] \to \mathbb{R}$ defined by $g(y) = f(-y)$, which is right continuous at $-c$ when f is left continuous at c; the verification of the appropriate relations among the indefinite integrals of f and g is straightforward but fussy.} ◊

9.3.5. Corollary. *If* $a < c < b$ *and* f *is continuous at* c, *then* F *and* H *are differentiable at* c *and* $F'(c) = H'(c) = f(c)$.

Proof. By assumption, f is both left and right continuous at c, so by 9.3.4 and 9.3.4',

$$F_l'(c) = f(c) = F_r'(c) \quad \text{and} \quad H_l'(c) = f(c) = H_r'(c) ;$$

by 8.1.2, F and H are differentiable at c, with $F'(c) = f(c)$ and $H'(c) = f(c)$. ◊

It is occasionally useful to regard the upper and lower integrals as functions of the lower endpoint of integration:

9.3.6. *Definition.* For the given bounded function $f : [a, b] \to \mathbb{R}$, we define functions $G : [a, b] \to \mathbb{R}$ and $K : [a, b] \to \mathbb{R}$ by the formulas

$$G(x) = \overline{\int_x^b} f \quad \text{and} \quad K(x) = \underline{\int_x^b} f$$

for $a \le x \le b$.

9.3.7. *Remarks.* In view of 9.2.3, we have

$$F(x) + G(x) = \overline{\int_a^b} f ,$$

$$H(x) + K(x) = \underline{\int_a^b} f ,$$

for $a \le x \le b$; thus G is in a sense *complementary* to F, and K to H. This is the key to deducing the properties of G from those of F, and the properties of K from those of H. For example, since F and H are continuous (9.3.2), so are G and K.

9.3.8. Theorem. *If* $a \le c < b$ *and* f *is right continuous at* c, *then* G *and* K *are right differentiable at* c *and* $G_r'(c) = K_r'(c) = -f(c)$.

Proof. This is immediate from 9.3.4 and the formulas in 9.3.7. ◊

Similarly,

9.3.8'. **Theorem.** *If $a < c \le b$ and f is left continuous at c, then G and K are left differentiable at c and $G'_l(c) = K'_l(c) = -f(c)$.*

9.3.9. **Corollary.** *If $a < c < b$ and f is continuous at c, then G and K are differentiable at c and $G'(c) = K'(c) = -f(c)$.*

Exercises

1. If $f \ge 0$ then the functions G and K of 9.3.6 are decreasing. If $f \le 0$ then G and K are increasing, while the functions F and H of 9.3.1 are decreasing.

2. If f is monotone then $F = H$ (and $G = K$). {Hint: §9.1, Exercise 4.}

3. If $a < c < b$ and

$$f(x) = \begin{cases} 0 & \text{for } a \le x < c \\ 1 & \text{for } c \le x \le b \end{cases}$$

then F is 'piecewise linear'; find its formula. {Hint: §9.2, Exercise 2.}

4. If $f(x) = x$ $(a \le x \le b)$, then $F(x) = \frac{1}{2}(x^2 - a^2)$. {Cf. §9.2, Exercise 3.}

5. In the Riemann-Stieltjes context (cf. §9.2, Exercise 4), upper and lower indefinite integrals are defined in the obvious way. In this context, the expression $k|x - y|$ in 9.3.2 gets replaced by $k|g(x) - g(y)|$; thus, the continuity properties of F hinge on those of g. The Riemann-Stieltjes analogue of 9.3.4: If f is right continuous at c, and g is right differentiable at c, then F and H are right differentiable at c with right derivative $f(c)g'_r(c)$.

6. The converses of 9.3.4–9.3.5 are false. {Cf. §9.6, Exercise 6.}

9.4. Riemann-Integrable Functions

9.4.1. **Definition.** A bounded function $f : [a, b] \to \mathbb{R}$ is said to be **Riemann-integrable** (briefly, *integrable*) if

$$\underline{\int_a^b} f = \overline{\int_a^b} f.$$

{The analogous concept for bounded sequences ($\lim\inf = \lim\sup$) is convergence!} We write simply

$$\int_a^b f\,,$$

or (especially when $f(x)$ is replaced by a formula for it)

$$\int_a^b f(x)dx\,,$$

for the common value of the lower and upper integral, and call it the **integral** (or *Riemann integral*) of f.

Most of this section is devoted to examples.

9.4.2. *Remark.* If f is Riemann-integrable then so is $-f$, and

$$\int_a^b (-f) = -\int_a^b f\,.$$

{This follows easily from 9.2.1.}

9.4.3. *Example.* If $f(x) = 1$ for x rational and $f(x) = 0$ for x irrational, then f is *not* Riemann-integrable (9.1.6).[1]

A very useful class of examples:

9.4.4. Theorem. *If f is monotone then it is Riemann-integrable.*

Proof. We can suppose that f is increasing (cf. 9.4.2). For every subdivision σ of $[a, b]$, we have

$$s(\sigma) \le \int_{\underline{a}}^b f \le \int_a^{\overline{b}} f \le S(\sigma)$$

(9.1.5, 9.1.13); to show that the lower integral is equal to the upper integral, we need only show that $S(\sigma) - s(\sigma)$ can be made as small as we like (by choosing σ appropriately). Say

$$\sigma = \{a = a_0 < a_1 < \cdots < a_n = b\}\,;$$

let us calculate explicitly the upper and lower sums for σ. In the notations of 9.1.2 and 9.1.3, we have

$$m_\nu = f(a_{\nu-1})\,, \quad M_\nu = f(a_\nu)$$

---

[1]There is a more comprehensive concept of integral, called the *Lebesgue integral*, for which this function *is* 'integrable', with integral 0 (cf. 11.5.4).

(because f is increasing), therefore

$$s(\sigma) = \sum_{\nu=1}^{n} f(a_{\nu-1})e_\nu, \quad S(\sigma) = \sum_{\nu=1}^{n} f(a_\nu)e_\nu$$

so

(*)
$$S(\sigma) - s(\sigma) = \sum_{\nu=1}^{n} [f(a_\nu) - f(a_{\nu-1})]e_\nu.$$

Now assume that the points of σ are equally spaced, so that

$$e_\nu = \frac{1}{n}(b-a) \qquad (\nu = 1, \ldots, n);$$

the sum in (*) then 'telescopes',

$$S(\sigma) - s(\sigma) = \frac{1}{n}(b-a) \sum_{\nu=1}^{n} [f(a_\nu) - f(a_{\nu-1})]$$

$$= \frac{1}{n}(b-a)[f(b) - f(a)],$$

which can be made arbitrarily small by taking n sufficiently large. ◊

The most famous (and useful!) example:

9.4.5. Theorem. *If f is continuous on $[a,b]$ then f is Riemann-integrable.*

Proof. Let F and H be the indefinite upper integral and indefinite lower integral defined in 9.3.1. We know that $F(a) = H(a) = 0$, and our problem is to show that $F(b) = H(b)$. By 9.3.2, F and H are continuous on $[a,b]$. By 9.3.5, we know that F and H are differentiable on (a,b) with $F'(x) = f(x) = H'(x)$ for all $x \in (a,b)$. Thus $F - H$ is continuous on $[a,b]$, differentiable on (a,b), and $(F-H)'(x) = 0$ for all $x \in (a,b)$, so $F - H$ is constant by a corollary (8.5.4) of the Mean Value Theorem; since $(F-H)(a) = 0$, also $(F-H)(b) = 0$, thus $F(b) = H(b)$ as we wished to show. ◊

This theorem and its proof are usually cast in the following form:

9.4.6. Theorem. (Fundamental Theorem of Calculus) *If $f : [a,b] \to \mathbb{R}$ is continuous, then*

(1) *f is Riemann-integrable on $[a,b]$;*

(2) *there exists a continuous function $F : [a,b] \to \mathbb{R}$, differentiable on (a,b), such that $F'(x) = f(x)$ for all $x \in (a,b)$;*

(3) *for any F satisfying (2), necessarily*

$$F(x) = F(a) + \int_a^x f \quad \text{for all } x \in [a,b];$$

moreover, F is right differentiable at a, left differentiable at b, and $F'_r(a) = f(a)$, $F'_l(b) = f(b)$.

Proof. (1) is the conclusion of 9.4.5. The function F introduced in the proof of 9.4.5 has the properties in (2) and (3); in particular, the properties at a and b follow from 9.3.4 and 9.3.4'.

Suppose that $J : [a, b] \to \mathbb{R}$ is also a continuous function having derivative $f(x)$ at every $x \in (a, b)$. By the argument in the proof of 9.4.5, $J - F$ is constant, say $J(x) = F(x) + C$ for all $x \in [a, b]$; then

$$J(x) - J(a) = F(x) - F(a) = \int_a^x f \quad \text{for all } x \in [a, b],$$

and J has the one-sided derivatives $f(a)$ and $f(b)$ at the endpoints (because F does). ◇

9.4.7. Corollary. *If* $f : [a, b] \to \mathbb{R}$ *is continuous and* $F : [a, b] \to \mathbb{R}$ *is a continuous function, differentiable on* (a, b), *such that* $F'(x) = f(x)$ *for all* $x \in (a, b)$, *then*

$$\int_a^b f = F(b) - F(a).$$

9.4.8. Corollary. *If* $f : [a, b] \to \mathbb{R}$ *is continuous,* $f \geq 0$ *on* $[a, b]$, *and*

$$\int_a^b f = 0,$$

then f *is identically* 0.

Proof. With F as in 9.4.6, F is increasing (9.3.3) and

$$F(b) - F(a) = \int_a^b f = 0,$$

therefore F is constant; then $f = F' = 0$ on (a, b), so $f = 0$ on $[a, b]$ by continuity. ◇

As we shall see in §9.6, considerably more can be squeezed out of the proof of 9.4.5; but first let's take a break from theory to harvest the application to logarithmic and exponential functions.

Exercises

1. If $a < c < b$, prove that f is Riemann-integrable on $[a, b]$ if and only if it is Riemann-integrable on both $[a, c]$ and $[c, b]$.
{Hint: For every closed subinterval $[c, d]$ of $[a, b]$, the number

$$\overline{\int_c^d} f - \underline{\int_c^d} f$$

is ≥ 0 and is an additive function of the endpoints (in the sense of 9.2.3); deduce that it is 0 for $[a,b]$ if and only if it is 0 for both $[a,c]$ and $[c,b]$.}

2. If f is Riemann-integrable and $c \in \mathbb{R}$ then cf is also Riemann-integrable and

$$\int_a^b cf = c \int_a^b f.$$

{Hint: §9.2, Exercise 1.}

3. Prove that if f and g are continuous on $[a,b]$, then

$$\int_a^b (f+g) = \int_a^b f + \int_a^b g.$$

{Hint: FTC. For the case of arbitrary Riemann-integrable functions, see 9.8.3.}

4. Corollary 9.4.8 has a much easier proof: assuming to the contrary that $f(c) > 0$ at some point c, argue that $f(x) \geq \frac{1}{2}f(c)$ on a subinterval of $[a,b]$ and construct a lower sum for f that is > 0.

5. True or false (explain): If $f \geq 0$ and

$$\overline{\int_a^b} f = 0,$$

then f is Riemann-integrable.

6. Call the bounded function $f : [a,b] \to \mathbb{R}$ *Riemann-Stieltjes integrable* (briefly, RS-integrable) with respect to the increasing function $g : [a,b] \to \mathbb{R}$ (cf. §9.1, Exercise 5) if the Riemann-Stieltjes lower and upper integrals of f with respect to g are equal; the common value is denoted $\int_a^b f\, dg$, called the *Riemann-Stieltjes integral* of f with respect to g.[2]

The analogue of 9.4.5 in this context: If f is continuous on $[a,b]$ then f is RS-integrable with respect to g.

{Hint: Let $\sigma = \{a = a_0 < a_1 < \cdots < a_n = b\}$ be a subdivision of $[a,b]$, calculate $S(\sigma) - s(\sigma)$, where, for example, $S(\sigma) = \sum_{\nu=1}^n M_\nu[g(a_\nu) - g(a_{\nu-1})]$, and use the uniform continuity of f (6.6.1) to choose σ so that $S(\sigma) - s(\sigma)$ is arbitrarily small.}

---

[2]This is one of several, not quite equivalent, variants of the 'Riemann-Stieltjes integral' [cf. W. Rudin, *Principles of mathematical analysis*, 3rd edn., McGraw-Hill, New York, 1976]. The intricacies of the subject are well sorted out in the books of T. H. Hildebrandt [*Introduction to the theory of integration*, p. 51, Theorem 10.9, Academic, New York, 1963] and K. A. Ross [*Elementary analysis: The theory of calculus*, §35, Springer-Verlag, New York, 1980].

*9.5. An Application: Log and Exp

The idea is to use a formula from 'elementary' calculus, together with
the machinery of the Riemann integral, to define in a rigorous way the
logarithmic and exponential functions. Your experience in calculus should
suffice for motivating the notations and definitions (you'll see a lot of old
friends wearing funny masks).

9.5.1. Definition. Define a function $L : (0, +\infty) \to \mathbb{R}$ by the formulas
(cf. Figure 14)

$$L(x) = \begin{cases} \displaystyle\int_1^x \frac{1}{t}\,dt & \text{for } x > 1 \\[2mm] 0 & \text{for } x = 1 \\[2mm] \displaystyle-\int_x^1 \frac{1}{t}\,dt & \text{for } 0 < x < 1. \end{cases}$$

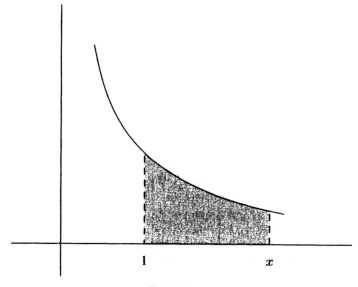

FIGURE 14

9.5.2. Convention. For an integrable function $f : [a, b] \to \mathbb{R}$ one writes

$$\int_b^a f = -\int_a^b f.$$

---

*This section can be omitted without loss of continuity in the rest of the
chapter.

(this is consistent with the computation of integrals via 'antiderivatives', as in 9.4.7). With this convention,

$$L(x) = \int_1^x \frac{1}{t}\,dt \quad \text{for all } x > 0 \qquad .$$

(even when the limits of integration are in the 'wrong order').

9.5.3. Theorem. *With L defined as in 9.5.1,*
   (i) $L(x) > 0 \Leftrightarrow x > 1$;
   (ii) $L(x) = 0 \Leftrightarrow x = 1$;
   (iii) $L(x) < 0 \Leftrightarrow 0 < x < 1$.
Moreover,
   (iv) $(x-1)/x \le L(x) \le x - 1$
for all $x > 0$.

Proof. (iv) For $x = 1$ the inequalities are trivial.
Suppose $x > 1$. For $1 \le t \le x$ we have

$$\frac{1}{x} \le \frac{1}{t} \le 1$$

therefore

$$\frac{1}{x}(x-1) \le \int_1^x \frac{1}{t}\,dt \le 1 \cdot (x-1)$$

by 9.1.8; thus (iv) holds for $x > 1$.
   Suppose $0 < x < 1$. For $x \le t \le 1$ we have

$$1 \le \frac{1}{t} \le \frac{1}{x},$$

therefore

$$1 \cdot (1-x) \le \int_x^1 \frac{1}{t}\,dt \le \frac{1}{x}(1-x)$$

by 9.1.8; multiplying by -1, we get the inequalities of (iv).
   (i)–(iii): The implications "$\Leftarrow$" are immediate from (iv), and the implications "$\Rightarrow$" follow from them by trichotomy in the set of positive real numbers (cf. the proof of 1.2.8). ◇

9.5.4. Theorem. $L'(x) = 1/x$ *for all $x > 0$.*

Proof. Fix $c > 0$; we have to show that L is differentiable at c and that $L'(c) = 1/c$.
   Choose real numbers a and b such that the closed interval $[a, b]$ contains both c and 1 as internal points. {For example, let $a = \frac{1}{2}c$ if $0 < c < 1$, and let $a = \frac{1}{2}$ if $c \ge 1$; in either case let $b = c+1$.} It will be convenient to apply the machinery of the preceding sections to the interval $[a, b]$ and the function $t \mapsto 1/t \ (a \le t \le b)$.

(i) Suppose $c > 1$. For x in a neighborhood of c, L is given by the formula

$$L(x) = \int_1^x \frac{1}{t} dt$$

with $x > 1$; by 9.3.5 (and 9.4.6) applied to the continuous function $t \mapsto 1/t$ on $[1, b]$, L is differentiable at c with $L'(c) = 1/c$, and by 9.3.4, L is right differentiable at 1 with $L_r'(1) = 1/1 = 1$.

(ii) Suppose $0 < c < 1$. For x in a neighborhood of c, L is given by the formula

$$L(x) = -\int_x^1 \frac{1}{t} dt$$

with $x < 1$; by 9.3.9 applied to $t \mapsto 1/t$ on the interval $[a, 1]$, L is differentiable at c with $L'(c) = -(-1/c) = 1/c$, and by 9.3.8′, L is left differentiable at 1 with $L_l'(1) = -(-1/1) = 1$.

(iii) Suppose $c = 1$. As noted in (i) and (ii), $L_l'(1) = L_r'(1) = 1$, so L is differentiable at 1 with $L'(1) = 1$ (8.1.2). $\Diamond$

9.5.5. Corollary. *The function $L : (0, +\infty) \to \mathbb{R}$ is strictly increasing.*

Proof. Since $L'(x) = 1/x > 0$ for all $x > 0$, this is immediate from the Mean Value Theorem (8.5.2). $\Diamond$

9.5.6. Theorem. $L(ab) = L(a) + L(b)$ *for all* $a > 0$, $b > 0$.

Proof. Fix $a > 0$ and define $f : (0, +\infty) \to \mathbb{R}$ by $f(x) = L(ax)$. Since $x \mapsto ax$ is differentiable at every $x > 0$, with derivative a, from 9.5.4 and the Chain Rule (8.3.1) we see that f is differentiable at x with

$$f'(x) = L'(ax) \cdot a = (1/ax)a = 1/x = L'(x);$$

thus $f - L$ is constant (8.5.4) and in particular

$$(f - L)(b) = (f - L)(1) = f(1) - L(1) = L(a) - 0,$$

so $L(a) = f(b) - L(b) = L(ab) - L(b)$. $\Diamond$

9.5.7. Corollary. $L(a^m) = mL(a)$ *for all* $a > 0$ *and all integers* m.

Proof. For a positive integer, this is an easy induction on m, since $L(a^{m+1}) = L(a^m a) = L(a^m) + L(a)$. Also, $L(a^{-1}) = -L(a)$ follows from applying L to $a^{-1}a = 1$, and the case of a negative integer follows from the preceding two cases. $\Diamond$

9.5.8. Corollary. *The function $L : (0, +\infty) \to \mathbb{R}$ is continuous and bijective.*

Proof. L is continuous (even differentiable) and injective (even strictly increasing); we have to show that every real number y belongs to the range of L. Since $L(2) > 0$ (9.5.3), by the Archimedean property we can choose positive integers m and n such that

$$mL(2) > y \quad \text{and} \quad nL(2) > -y.$$

Then

$$(-n)L(2) < y < mL(2),$$

that is (9.5.7),

$$L(2^{-n}) < y < L(2^m),$$

so y belongs to the range of L by the Intermediate Value Theorem (6.1.2). $\Diamond$

9.5.9. *Definition.* We write $E = L^{-1}$ for the inverse of L.

Thus $E : \mathbb{R} \to (0, +\infty)$ is bijective, continuous (6.5.2) and strictly increasing (by 9.5.5). The properties verified for L are reflected in its inverse:

(i) $E(x) > 1 \Leftrightarrow x > 0$;

(ii) $E(x) = 1 \Leftrightarrow x = 0$;

(iii) $0 < E(x) < 1 \Leftrightarrow x < 0$;

(iv) $E(x + y) = E(x)E(y)$ for all x, y in $\mathbb{R}$;

(v) $a^m = E(mL(a))$ for all $a > 0$ and $m \in \mathbb{Z}$.

{For example, to verify the equations (iv) and (v), it suffices to apply L to both sides and observe that the results are equal.}

On the way to proving that $E' = E$:

9.5.10. *Lemma.*

$$\lim_{x \to 1} \frac{L(x)}{x - 1} = 1.$$

Proof. {At $x = 1$, $L(x)$ and $x - 1$ are both 0; so we are looking at an 'indeterminate form'.} The lemma follows from the inequalities (iv) in 9.5.3: dividing by $x - 1$ (with or without reversal of inequalities), the fraction in question is caught between $1/x$ and 1. $\Diamond$

9.5.11. *Lemma.*

$$\lim_{h \to 0} \frac{E(h) - 1}{h} = 1.$$

Proof. Let $h_n \to 0$, $h_n \neq 0$. Since E is continuous, $E(h_n) \to E(0) = 1$. Let $x_n = E(h_n)$; then $x_n > 0$, $x_n \to 1$ and $x_n \neq 1$ (because $h_n \neq 0$), so by 9.5.10

$$\frac{L(x_n)}{x_n - 1} \to 1,$$

that is,

$$\frac{h_n}{E(h_n) - 1} \to 1,$$

whence the lemma. $\Diamond$

9.5.12. Theorem. $E' = E$.

Proof. Let $x \in \mathbb{R}$; we have to show that E is differentiable at x, with $E'(x) = E(x)$. If $h \neq 0$ then

$$\frac{E(x+h) - E(x)}{h} = \frac{E(x)E(h) - E(x)}{h}$$

$$= E(x) \cdot \frac{E(h) - 1}{h} \, ;$$

by 9.5.11, this tends to $E(x)$ as $h \to 0$, whence the theorem. $\Diamond$

The function E is the key to a general concept of exponential:

9.5.13. *Definition.* If $a > 0$ and $x \in \mathbb{R}$, we define

$$a^x = E(xL(a)) \, .$$

{For $x \in \mathbb{Z}$ this agrees with the elementary definition of a^x, by item (v) following 9.5.9.} Note that

$$L(a^x) = xL(a)$$

(because $L = E^{-1}$).

9.5.14. Theorem. *For all* $a > 0$, $b > 0$ *and all* x, y *in* $\mathbb{R}$,
(1) $a^{x+y} = a^x a^y$,
(2) $(a^x)^y = a^{xy}$,
(3) $(ab)^x = a^x b^x$.

Proof. (1) By definition, $a^{x+y} = E((x+y)L(a))$, whereas

$$a^x a^y = E(xL(a)) \cdot E(yL(a)) = E(xL(a) + yL(a)) \, ;$$

so it all comes down to the distributive law in $\mathbb{R}$.
(2) Note that $a^x > 0$, so the left side of (2) makes sense; and

$$(a^x)^y = E(yL(a^x)) = E(y \cdot xL(a)) \, ,$$

whereas $a^{xy} = E(xy \cdot L(a))$; cite the associative and commutative laws for multiplication in $\mathbb{R}$.
(3) $(ab)^x = E(xL(ab)) = E(x[L(a) + L(b)])$, whereas

$$a^x b^x = E(xL(a)) \cdot E(xL(b)) = E(xL(a) + xL(b)) \, ;$$

tally another point for the distributive law. $\Diamond$

9.5.15. Corollary. *If* $a > 0$, m *and* n *are integers, and* $n \neq 0$, *then*

$$(a^{1/n})^n = a,$$
$$a^{m/n} = (a^{1/n})^m = (a^m)^{1/n}.$$

Proof. By 9.5.14, $(a^{1/n})^n = a^{(1/n)n} = a^1 = a$, and similarly for the other equations. {When n is a positive integer, the first equation shows that $a^{1/n}$ is a roundabout way of constructing n'th roots (cf. 6.2.2); the message of the other equations is that rational exponents have the desired interpretation in terms of roots and integral powers.} ◊

Finally, we introduce the familiar notations for the exponential and logarithmic functions of calculus:

9.5.16. *Definition.* $e = E(1)$.
Then $L(e) = 1 > 0$, so $e > 1$.

9.5.17. *Theorem.* (1) $E(x) = e^x$ *for all* $x \in \mathbb{R}$;
(2) $L(e^x) = x$ *for all* $x \in \mathbb{R}$;
(3) *for* $y > 0$ *and* $x \in \mathbb{R}$, $L(y) = x$ *if and only if* $y = e^x$.

Proof. (1) $e^x = E(xL(e)) = E(x \cdot 1)$.
(2), (3) are immediate from (1) and the fact that E and L are each other's inverses; in the classical notation,

$$L(y) = \log_e y = \ln y$$

(the 'natural' or Naperian logarithm of y). ◊

On the way to a limit formula for e^x:

9.5.18. *Lemma.* *If* $h > -1$ *then*

$$\frac{h}{1+h} \leq L(1+h) \leq h.$$

Proof. Put $x = 1 + h$ in (iv) of 9.5.3. ◊

9.5.19. *Theorem.* *For all* $x \in \mathbb{R}$,

$$e^x = \lim_{n \to \infty} \left(1 + \frac{x}{n}\right)^n.$$

Proof. The formula is trivial for $x = 0$. Assuming $x \neq 0$, let $h_n = x/n$ and let $a_n = (1 + h_n)^n$; then $n = x/h_n$ and, for n sufficiently large, $a_n > 0$ and

$$L(a_n) = nL(1 + h_n) = x \cdot \frac{L(1 + h_n)}{h_n}.$$

Since $h_n \to 0$, $L(a_n) \to x$ by 9.5.10 (consider $x_n = 1 + h_n \to 1$), therefore

$$a_n = E(L(a_n)) \to E(x) = e^x$$

by the continuity of E. $\Diamond$

In particular,

$$e = \lim_{n \to \infty} \left(1 + \frac{1}{n}\right)^n .$$

Exercises

1. Prove that a^x is 'jointly continuous in a and x', in the following sense: if $a_n > 0$, $a_n \to a > 0$ and $x_n \to x$ in $\mathbb{R}$, then $a_n^{x_n} \to a^x$.

2. Fix $c \in \mathbb{R}$ and define $f : (0, +\infty) \to (0, +\infty)$ by $f(x) = x^c$. Prove that f is differentiable and $f'(x) = cx^{c-1}$ for all $x > 0$. {Hint: Look at $\ln \circ f$.}

3. Fix $c > 0$ and define $g : \mathbb{R} \to (0, +\infty)$ by $g(x) = c^x$. Prove that g is differentiable and $g'(x) = (\ln c)c^x$.

4. This exercise sketches the initial steps of a conceivable (albeit clumsy) development of the trigonometric functions.

Let $f : \mathbb{R} \to (0, 1]$ be the function $f(x) = 1/(1 + x^2)$. Define $A : \mathbb{R} \to \mathbb{R}$ by the formulas

$$A(x) = \int_0^x f \quad \text{for } x \geq 0,$$

$$A(x) = -\int_x^0 f \quad \text{for } x < 0.$$

(The letter A anticipates the arctangent function.)
  (i) Show that A is an odd function: $A(-x) = -A(x)$ for all $x \in \mathbb{R}$. {Hint: f is an even function.}
  (ii) A is strictly increasing. {Hint: $A' = f$.}
  (iii) For every positive integer k,

$$\frac{1}{1 + k^2} \leq A(k) - A(k - 1) \leq \frac{1}{1 + (k - 1)^2} .$$

{Hint: Integrate f over the interval $[k - 1, k]$.}
  (iv) Let

$$s_n = \sum_{k=1}^n \frac{1}{1 + k^2} .$$

Deduce from (iii) that $s_n \leq A(n) \leq 1 + s_{n-1}$ for every positive integer n, consequently A is bounded. {Hint: It is not difficult to show that

$$\sum_{k=1}^{n} \frac{1}{k^2} < 2$$

for every positive integer n (10.3.4).}

(v) The range of A is an open interval $(-a, a)$. {Hint: (i), (ii), (iv) and the IVT.} Define $\pi = 2a$ (but we're a long way from the circumference of a circle!).

(vi) Let $T : (-a, a) \to \mathbb{R}$ be the inverse function $T = A^{-1}$ (also strictly increasing and odd). Show that $T' = 1 + T^2$. {Hint: §8.5, Exercise 2.} The notation T anticipates the tangent function on $(-\pi/2, \pi/2)$.}

(vii) Define $C : (-a, a) \to \mathbb{R}$ by the formula

$$C = \frac{1}{(1 + T^2)^{1/2}} = \frac{1}{(T')^{1/2}}.$$

Thus $1/C^2 = 1 + T^2$. {The notation C anticipates the cosine function; remember $\sec^2 = 1 + \tan^2$?}

(viii) Define $S : (-a, a) \to \mathbb{R}$ by $S = TC$ and note that $S^2 + C^2 = 1$, $C' = -S$, $S' = C$.

5. Prove that $(\ln n)/n \to 0$ and infer that $n^{1/n} \to 1$. {Hint: If $t \geq 1$ then $1/t \leq 1/t^{1/2}$, whence $\ln n \leq 2(n^{1/2} - 1)$.}

9.6. Piecewise Pleasant Functions

We take up again the notations established at the beginning of the chapter. Our objective: to enlarge substantially the class of functions we know to be Riemann-integrable.

The concept of Riemann-integrability is 'additive' in the following sense:

9.6.1. **Theorem.** *Let $a < c < b$. The following conditions on a bounded function $f : [a, b] \to \mathbb{R}$ are equivalent:*
(a) *f is Riemann-integrable on $[a, b]$;*
(b) *f is Riemann-integrable on both $[a, c]$ and $[c, b]$.*

Proof. {The meaning of (b) is that the restricted functions $f|[a, c]$ and $f|[c, b]$ are Riemann-integrable.} Let

$$A = \overline{\int_a^b} f - \underline{\int_a^b} f,$$

$$B = \overline{\int_a^c} f - \underline{\int_a^c} f,$$

$$C = \overline{\int_c^b} f - \underline{\int_c^b} f.$$

Then A, B, C are ≥ 0 (9.1.13) and it follows from 9.2.3 that $A = B+C$, therefore $A = 0$ if and only if $B = C = 0$. $\Diamond$

9.6.2. Corollary. *Let*

$$\sigma = \{a = a_0 < a_1 < \cdots < a_n = b\}$$

be a subdivision of $[a, b]$*;* f *is Riemann-integrable on* $[a, b]$ *if and only if it is Riemann-integrable on every subinterval* $[a_{\nu-1}, a_\nu]$ *(*$\nu = 1, \ldots, n$*).*

9.6.3. Theorem. *Suppose* $f : [a, b] \to \mathbb{R}$ *is a bounded function such that, whenever* $a < c < d < b$*,* f *is Riemann-integrable on* $[c, d]$*. Then* f *is Riemann-integrable on* $[a, b]$*.*

Proof. With notations as in 9.3.1, let F and H be the indefinite upper and lower integrals of f. If $a < c < d < b$ then, by 9.2.3, we have

$$F(d) = F(c) + \int_c^{\overline{d}} f$$

and

$$H(d) = H(c) + \int_{\underline{c}}^d f\,;$$

the last terms on the right are equal by hypothesis, so

$$F(d) - F(c) = H(d) - H(c).$$

Let (c_n) and (d_n) be sequences such that

$$a < c_n < d_n < b$$

and $c_n \to a$, $d_n \to b$; then

$$F(d_n) - F(c_n) = H(d_n) - H(c_n)$$

for all n, so in the limit we have

$$F(b) - 0 = H(b) - 0$$

by the continuity of F and H (9.3.2). In other words,

$$\int_a^{\overline{b}} f = \int_{\underline{a}}^b f. \ \Diamond$$

9.6.4. Definition. We say that $f : [a, b] \to \mathbb{R}$ is **piecewise monotone** if there exists a subdivision $\sigma = \{a = a_0 < a_1 < \cdots < a_n = b\}$ of

$[a, b]$ such that f is monotone on each of the open intervals $(a_{\nu-1}, a_\nu)$
$(\nu = 1, \ldots, n)$.

9.6.5. **Theorem.** *If the bounded function* $f : [a, b] \to \mathbb{R}$ *is piecewise monotone, then it is Riemann-integrable.*

Proof. In view of 9.6.2, we can suppose f is monotone on (a, b). If $a < c < d < b$ then f is monotone on $[c, d]$, therefore Riemann-integrable on $[c, d]$ (9.4.4); by 9.6.3, f is Riemann-integrable on $[a, b]$. $\Diamond$

9.6.6. **Definition.** We say that $f : [a, b] \to \mathbb{R}$ is **piecewise continuous** if there exists a subdivision $\sigma = \{a = a_0 < a_1 < \cdots < a_n = b\}$ of $[a, b]$ such that f is continuous on each of the open subintervals $(a_{\nu-1}, a_\nu)$ $(\nu = 1, \ldots, n)$. {It is the same to say that f is continuous at all but finitely many points of $[a, b]$.}

9.6.7. **Theorem.** *If the bounded function* $f : [a, b] \to \mathbb{R}$ *is piecewise continuous, then it is Riemann-integrable.*

Proof. In view of 9.6.2, we can suppose f is continuous on (a, b). If $a < c < d < b$ then f is continuous on $[c, d]$, therefore Riemann-integrable on $[c, d]$ (9.4.5); by 9.6.3, f is Riemann-integrable on $[a, b]$. $\Diamond$

We now turn to a different technique for proving Riemann-integrability.

9.6.8. **Definition.** If $f : X \to Y$ and $g : X \to Y$ are functions such that the set

$$\{x \in X : f(x) \neq g(x)\}$$

is *finite* (possibly empty), in other words if $f(x) = g(x)$ with only *finitely many exceptions* (possibly none), we write

$$f = g \quad \text{f.e.}$$

Another way to express it: f and g agree at all but finitely many points of their common domain.

In particular, for functions $f : [a, b] \to \mathbb{R}$ and $g : [a, b] \to \mathbb{R}$, to say that $f = g$ f.e. means that there exists a subdivision $\sigma = \{a = a_0 < a_1 < \cdots < a_n = b\}$ of $[a, b]$ such that $f = g$ on each of the open subintervals $(a_{\nu-1}, a_\nu)$.

Changing a function at finitely many points has no effect on upper and lower integrals:

9.6.9. **Theorem.** *If* $f : [a, b] \to \mathbb{R}$ *and* $g : [a, b] \to \mathbb{R}$ *are bounded functions such that* $f = g$ *f.e., then*

$$\overline{\int_a^b} f = \overline{\int_a^b} g \quad and \quad \underline{\int_a^b} f = \underline{\int_a^b} g.$$

Proof. We can restrict attention to upper integrals (9.2.1). In view of 9.2.3, we can suppose that $f(x) = g(x)$ for all $x \in (a, b)$.

Let $F : [a, b] \to \mathbb{R}$ and $G : [a, b] \to \mathbb{R}$ be the indefinite upper integrals of f and g, respectively:

$$F(x) = \overline{\int_a^x} f \quad \text{and} \quad G(x) = \overline{\int_a^x} g$$

for $a \leq x \leq b$. If $a < c < d < b$ then, as in the proof of 9.6.3,

$$F(d) = F(c) + \overline{\int_c^d} f \,,$$

$$G(d) = G(c) + \overline{\int_c^d} g \,,$$

and, since $f = g$ on $[c, d]$, we have

$$F(d) - F(c) = G(d) - G(c) \,;$$

as in the proof of 9.6.3, $F(b) = G(b)$ follows from the continuity of F and G. $\Diamond$

9.6.10. **Corollary.** *If $f : [a, b] \to \mathbb{R}$ and $g : [a, b] \to \mathbb{R}$ are functions such that g is Riemann-integrable and $f = g$ f.e., then f is also Riemann-integrable and*

$$\int_a^b f = \int_a^b g \,.$$

Proof. This is immediate from 9.6.9. $\Diamond$

9.6.11. **Theorem.** *If the bounded function $f : [a, b] \to \mathbb{R}$ has a right limit at every point of the open interval (a, b), then f is Riemann-integrable.*

Proof. Let F and H be as in the proof of 9.6.3. Given $a < c < b$, let us show that F and H are right differentiable at c and $F_r'(c) = H_r'(c)$. Let $g : [a, b] \to \mathbb{R}$ be the function such that

$$g(x) = f(x) \quad \text{for all } x \in [a, b] - \{c\} \,,$$
$$g(c) = f(c+) \,.$$

Thus g is right continuous at c and, by 9.6.9, the indefinite upper and lower integrals for g are the same as those for f, namely F and H. Since g is right continuous at c, it follows that F and H are right differentiable at c with $F_r'(c) = g(c) = H_r'(c)$ (9.3.4). This proves the assertion in the first paragraph.

Forget the preceding paragraph; it has done its duty. We now know that $F - H$ is continuous on $[a, b]$ (9.3.2), right differentiable on (a, b), and

$(F - H)'_r(c) = 0$ for all $c \in (a, b)$; by 8.5.5, $F - H$ is a constant function. In particular,
$$(F - H)(b) = (F - H)(a) = 0 - 0,$$
so $F(b) = H(b)$; in other words, f is Riemann-integrable.[1] $\Diamond$

9.6.12. Corollary. *If the bounded function* $f : [a, b] \to \mathbb{R}$ *has a right limit at all but finitely many points of* $[a, b]$, *then* f *is Riemann-integrable.*

Proof. This is immediate from 9.6.2 and 9.6.11. $\Diamond$

Note that this corollary includes 9.6.5 and 9.6.7 as special cases (but its proof is much harder, since it relies on 8.5.5).

Exercises

1. Let $g : [a, b] \to \mathbb{R}$ be Riemann-integrable, $f : [a, b] \to \mathbb{R}$ a bounded function, (x_n) a sequence of points in $[a, b]$ such that $f(x) = g(x)$ for all x in $[a, b]$ other than the x_n. Give an example to show that f need not be Riemann-integrable. In other words, the conclusion of 9.6.10 may be false if the finite number of exceptional points is replaced by a sequence of exceptional points.

2. Let $f : [c, +\infty) \to \mathbb{R}$ and assume that $f|[c, x]$ is integrable for every $x > c$. We say that f is *integrable over* $[c, +\infty)$ if the integral of $f|[c, x]$ has a limit L as $x \to +\infty$, in the following sense: for every $\epsilon > 0$, there exists an $\eta > 0$ such that
$$x > \eta \; \Rightarrow \; \left| \int_c^x f - L \right| < \epsilon.$$

Such an L is then unique; it is called the *integral* of f over $[c, +\infty)$ and is denoted
$$\int_c^{+\infty} f \quad \text{or} \quad \int_c^{+\infty} f(x)dx.$$

With f as in the first sentence of the exercise, suppose $c < d < +\infty$. Prove that f is integrable over $[c, +\infty)$ if and only if it is integrable over $[d, +\infty)$, in which case
$$\int_c^{+\infty} f = \int_c^d f + \int_d^{+\infty} f.$$

3. State and prove the analogue of Exercise 2 for a function $f : (-\infty, c] \to \mathbb{R}$.

---

[1] I learned this ingenious proof from a paper by R. Metzler [*Amer. Math. Monthly* **78** (1971), 1129–1131].

4. Let $f : \mathbb{R} \to \mathbb{R}$ and assume that $f|[x, y]$ is integrable for every closed interval $[x, y]$ of $\mathbb{R}$. If there exists a point $c \in \mathbb{R}$ such that the integrals

$$\int_{-\infty}^{c} f \quad \text{and} \quad \int_{c}^{+\infty} f$$

exist in the sense of Exercises 2 and 3, then they exist for *every* $c \in \mathbb{R}$ and the sum

$$\int_{-\infty}^{c} f + \int_{c}^{+\infty} f$$

is independent of c; this sum is called the *integral* of f (over $\mathbb{R}$) and is denoted

$$\int_{-\infty}^{+\infty} f .$$

5. Let $f : (a, b) \to \mathbb{R}$ and assume that $f|[x, y]$ is integrable for every closed subinterval $[x, y]$ of (a, b). With an eye on Exercises 2–4, discuss the concept of integrability of f over (a, b). (Integrals of the type discussed in Exercises 2–5 are called *improper*.)

6. If a bounded function on $[a, b]$ has a right limit at a point $c \in [a, b)$, then its indefinite upper and lower integrals are right differentiable at c (see the proof of 9.6.11). The converse of this proposition is false, as the following example shows.
  Let $A = \{1 - 1/n : n \in \mathbb{P}\} \cup \{1 + 1/n : n \in \mathbb{P}\}$ and let $f : [0, 2] \to \mathbb{R}$ be the function defined by

$$f(x) = \begin{cases} 1 & \text{if } x \in A \\ 0 & \text{if } x \notin A . \end{cases}$$

  (i) Sketch the graph of f.
  (ii) f is Riemann-integrable.
  (iii) The indefinite integral F of f is identically zero.
  (iv) $F'(1) = f(1)$, but f has neither a left limit nor a right limit at 1.
  {Hint: (ii) 9.6.12. (iii) First argue that F is zero on $[0, 1 - 1/n]$ and on $[1 + 1/n, 2]$.}

7. Let $f : [a, b] \to \mathbb{R}$ be a bounded function, $N = \{c_n : n \in \mathbb{P}\}$ for some sequence (c_n) in $[a, b]$. Prove: If f has a right limit at every point of $(a, b) - N$, then f is Riemann-integrable.
  {Hint: If F and H are the upper and lower indefinite integrals of f, apply §8.5, Exercise 10 to the function $F - H$.}

8. The function $f : [0, 1] \to [0, 1]$ of §5.2, Exercise 6 ($f(x) = 0$ if $x = 0$ or x is irrational, $f(x) = 1/n$ if $x = m/n$ is a nonzero rational in reduced form) is Riemann-integrable, with integral zero. {Hint: Exercise 7.}

9. Explore the analogues of the results of this section for the Riemann-Stieltjes integral $\int_a^b f\,dg$ defined in §9.4, Exercise 6. The facts are as follows.

(i) The analogues of 9.6.1 and 9.6.2 hold with no further restrictions on the increasing function g.

(ii) The analogues of 9.6.3, 9.6.7, 9.6.9–9.6.10 hold assuming g is continuous on $[a,b]$.

(iii) The analogues of 9.6.5 and 9.6.11–9.6.12 hold assuming g is continuous on $[a,b]$ and right differentiable on (a,b) (and the latter condition can be relaxed to allow a sequence of exceptional points).

For example, the following result can be squeezed out of the proof of Theorem 9.6.11: If the bounded function $f : [a,b] \to \mathbb{R}$ has a right limit at all but a sequence of points of the open interval (a,b), g is continuous on the closed interval $[a,b]$, and g has a right derivative at all but a sequence of points of (a,b), then f is RS-integrable with respect to g and the indefinite RS-integral $x \mapsto \int_a^x f\,dg$ has a right derivative equal to $f(x+)g_r'(x)$ at all but a sequence of points of (a,b). {Hint: §9.3, Exercise 5 and §8.5, Exercise 10.}

9.7. Darboux's Theorem

The rest of the chapter is modeled on E. Landau's superb exposition.[1]

9.7.1. *Definition.* With notations as in 9.1.1 and 9.1.2, recall that the number $\omega_\nu = M_\nu - m_\nu$ is called the **oscillation** of f over the subinterval $[a_{\nu-1}, a_\nu]$. We also write

$$W_f(\sigma) = S(\sigma) - s(\sigma) = \sum_{\nu=1}^n \omega_\nu e_\nu = \sum_{\nu=1}^n (M_\nu - m_\nu)(a_\nu - a_{\nu-1}),$$

called the **weighted oscillation** of f for the subdivision σ of $[a,b]$. {The 'weights' are the subinterval lengths e_ν; the significance of the oscillation ω_ν is tempered by the length of the subinterval over which it takes place. Note that the weighted oscillation incorporates data on both the swing in y (via the ω_ν) and the swing in x (via the e_ν).}

9.7.2. *Remark.* If $\tau \succ \sigma$ then $W_f(\tau) \le W_f(\sigma)$. {This is immediate from 9.1.11.}

9.7.3. *Lemma.* *If σ and τ are any subdivisions of $[a,b]$ and ρ is a common refinement of σ and τ, then*

$$W_f(\rho) \le S(\tau) - s(\sigma).$$

---

[1] E. Landau, *Differential and integral calculus* [Chelsea, New York, 1951]; my nominee for the best calculus book ever written.

Proof. This is immediate from the chain of inequalities in the proof of 9.1.12. ◊

9.7.4. **Lemma.** *For every bounded function* $f : [a, b] \to \mathbb{R}$,

$$\inf_\sigma W_f(\sigma) = \overline{\int_a^b} f - \underline{\int_a^b} f .$$

Proof. Write W for the infimum on the left side; since $W_f(\sigma) \geq 0$ for all σ, we have $W \geq 0$.

From the definitions (9.1.5) it is immediate that

$$W_f(\sigma) \geq \overline{\int_a^b} f - \underline{\int_a^b} f$$

for every subdivision σ, therefore

$$W \geq \overline{\int_a^b} f - \underline{\int_a^b} f .$$

To establish the reverse inequality, choose sequences (σ_n) and (τ_n) of subdivisions of $[a, b]$ such that

$$s(\sigma_n) \to \underline{\int_a^b} f \quad \text{and} \quad S(\tau_n) \to \overline{\int_a^b} f$$

(possible by the definitions of upper and lower integrals). For each n, let ρ_n be a common refinement of σ_n and τ_n; citing 9.7.3, we have

$$W \leq W_f(\rho_n) \leq S(\tau_n) - s(\sigma_n)$$

and passage to the limit yields

$$W \leq \overline{\int_a^b} f - \underline{\int_a^b} f$$

by 3.4.8, (8). ◊

9.7.5. **Theorem.** *The following conditions on a bounded function* $f : [a, b] \to \mathbb{R}$ *are equivalent:*

(a) f *is Riemann-integrable;*

(b) *for every* $\epsilon > 0$, *there exists a subdivision* σ *of* $[a, b]$ *such that* $W_f(\sigma) \leq \epsilon$; *in other words,*

$$\inf_\sigma W_f(\sigma) = 0 .$$

Proof. This is immediate from 9.7.4 and the definition of Riemann-integrability. ◊

The process of perfecting approximations by 'indefinite refinement' has the flavor of passing to a limit; the thrust of the preceding remarks is that f is Riemann-integrable if and only if the weighted oscillation tends to 0 as σ ... as σ ... shall we say "$\sigma \to \infty$"? There is another way to 'get to ∞': by making $N(\sigma)$—the norm of the subdivision σ (9.1.1)—tend to 0. There's a technical problem: if $\tau \succ \sigma$ then $N(\tau) \leq N(\sigma)$; but $N(\tau) \leq N(\sigma)$ does not imply that τ refines σ (apart from a and b, they may have no points in common!). This complicates the formulation of Riemann-integrability in terms of a limit as $N(\sigma) \to 0$, and it is this technical difficulty that Darboux's theorem overcomes (cf. 9.7.10).

9.7.6. Lemma. *Let σ be any subdivision of $[a, b]$, let $a < c < b$ and let τ be the refinement of σ obtained by inserting the point c. If $|f(x)| \leq K$ for all $x \in [a, b]$, then*

$$S(\tau) \geq S(\sigma) - 2K \cdot N(\sigma).$$

Proof. What does the lemma say? We know that refinement can only shrink an upper sum; the inequality says that insertion of a point can't shrink the upper sum by more than $2K \cdot N(\sigma)$ (see Figure 15).

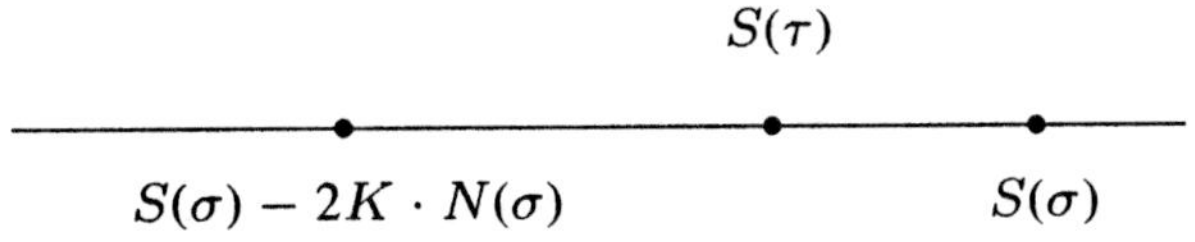

$$S(\tau)$$

$$S(\sigma) - 2K \cdot N(\sigma) \qquad\qquad S(\sigma)$$

FIGURE 15

Say $\sigma = \{a = a_0 < a_1 < \cdots < a_n = b\}$. If $c = a_\nu$ for some ν, then $\tau = \sigma$ and the inequality is trivial. Suppose $a_{\mu-1} < c < a_\mu$. Let

$$M' = \sup\{f(x): \ a_{\mu-1} \leq x \leq c\},$$
$$M'' = \sup\{f(x): \ c \leq x \leq a_\mu\}.$$

As noted in the proof of 9.1.11,

$$S(\sigma) - S(\tau) = M_\mu e_\mu - [M'(c - a_{\mu-1}) + M''(a_\mu - c)]$$
$$= (M_\mu - M')(c - a_{\mu-1}) + (M_\mu - M'')(a_\mu - c);$$

since M_μ, M', M'' all belong to $[-K, K]$, so that

$$|M_\mu - M'| \leq 2K, \quad |M_\mu - M''| \leq 2K,$$

we have

$$S(\sigma) - S(\tau) \le 2K(c - a_{\mu-1}) + 2K(a_\mu - c)$$
$$= 2K(a_\mu - a_{\mu-1}) = 2Ke_\mu \le 2K \cdot N(\sigma). \ \Diamond$$

More generally,

9.7.7. Lemma. *If τ is obtained from σ by inserting r new points, and if $|f| \le K$ on $[a,b]$, then*

$$S(\tau) \ge S(\sigma) - 2rK \cdot N(\sigma).$$

Proof. Let $\tau = \tau_r \succ \tau_{r-1} \succ \ldots \succ \tau_1 \succ \tau_0 = \sigma$, where τ_i is obtained from τ_{i-1} by inserting one new point. By the preceding lemma,

$$S(\tau_{i-1}) - S(\tau_i) \le 2K \cdot N(\tau_{i-1}) \le 2K \cdot N(\sigma)$$

for $i = 1,\ldots,r$; summing these inequalities over i, the left members telescope to $S(\sigma) - S(\tau)$ and the right members add up to $2Kr \cdot N(\sigma)$. $\Diamond$

9.7.8. Theorem. (Darboux's theorem[2]) *For every bounded function $f : [a,b] \to \mathbb{R}$,*

$$\lim_{N(\sigma)\to 0} S_f(\sigma) = \overline{\int_a^b} f$$

in the following sense: for every $\epsilon > 0$, there exists a $\delta > 0$ such that, for subdivisions σ of $[a,b]$,

$$N(\sigma) \le \delta \ \Rightarrow \ \left| S_f(\sigma) - \overline{\int_a^b} f \right| \le \epsilon.$$

Proof. Since

$$\overline{\int_a^b} f \le S(\sigma)$$

for every upper sum $S(\sigma) = S_f(\sigma)$, the inequality on the right side of the implication is equivalent to

(*) $$S(\sigma) \le \overline{\int_a^b} f + \epsilon.$$

Let $K > 0$ be such that $|f| \le K$ on $[a,b]$. Given any $\epsilon > 0$, choose a subdivision σ_0 of $[a,b]$ such that

$$S(\sigma_0) \le \overline{\int_a^b} f + \epsilon/2$$

---

[2]Published in 1875 independently by K. J. Thomae, G. Ascoli, H. J. S. Smith and G. Darboux [cf. T. Hawkins, *Lebesgue's theory of integration: Its origins and development*, 2nd edn., pp. 40-41, Chelsea, 1975].

(possible by the definition of upper integral as a greatest lower bound). Say σ_0 has m points other than the endpoints a and b; refining σ_0 if necessary, we can suppose $m \geq 1$. Let $\delta = \epsilon/4mK$. Assuming σ is any subdivision of $[a, b]$ with $N(\sigma) \leq \delta$, let us verify the inequality (*).

Let τ be the 'least common refinement' of σ_0 and σ (τ is the result of lumping together the points of σ_0 and σ). Let r be the number of points of τ that are not in σ (conceivably $\tau = \sigma$, $r = 0$); these points come from σ_0, so $0 \leq r \leq m$. By 9.7.7,

$$S(\tau) \geq S(\sigma) - 2rK \cdot N(\sigma);$$

but

$$2rK \cdot N(\sigma) \leq 2mK\delta = \epsilon/2,$$

so

(i) $$S(\tau) \geq S(\sigma) - \epsilon/2;$$

also $\tau \succ \sigma_0$, so

(ii) $$S(\tau) \leq S(\sigma_0) \leq \overline{\int_a^b} f + \epsilon/2,$$

and (*) results from (i) and (ii). ◊

The analog of 9.7.8 for lower integrals:

9.7.8′. Theorem. *For every bounded function* $f : [a, b] \to \mathbb{R}$,

$$\lim_{N(\sigma) \to 0} s_f(\sigma) = \underline{\int_a^b} f$$

in the following sense: for every $\epsilon > 0$, *there exists a* $\delta > 0$ *such that, for subdivisions* σ *of* $[a, b]$,

$$N(\sigma) \leq \delta \;\Rightarrow\; \left| s_f(\sigma) - \underline{\int_a^b} f \right| \leq \epsilon.$$

Proof. Since

$$s_f(\sigma) = -S_{-f}(\sigma),$$

this is immediate from 9.7.8 (applied to $-f$) and 9.2.1. ◊

Combining 9.7.8 and 9.7.8′,

9.7.8″. Theorem. *Let* $f : [a, b] \to \mathbb{R}$ *be any bounded function. For every* $\epsilon > 0$ *there exists a* $\delta > 0$ *such that, for every pair of subdivisions* σ *and* τ *of* $[a, b]$ *with* $N(\sigma) \leq \delta$ *and* $N(\tau) \leq \delta$, *we have*

$$\underline{\int_a^b} f - \epsilon \leq s_f(\sigma) \leq S_f(\tau) \leq \overline{\int_a^b} f + \epsilon,$$

and in particular (letting $\tau = \sigma$ *)*

$$N(\sigma) \leq \delta \quad \Rightarrow \quad W_f(\sigma) \leq \int_a^{\overline{b}} f - \int_{\underline{a}}^b f + 2\epsilon.$$

From this we deduce a 'norm' variant of 9.7.4:

9.7.9. Lemma. *For every bounded function* $f : [a, b] \to \mathbb{R}$,

$$\lim_{N(\sigma) \to 0} W_f(\sigma) = \int_a^{\overline{b}} f - \int_{\underline{a}}^b f.$$

Proof. By 9.7.4, the expression on the right side is equal to the infimum W of all the weighted oscillations. Given any $\epsilon > 0$, choose $\delta > 0$ as in 9.7.8''; if σ is any subdivision with $N(\sigma) \leq \delta$, then $W \leq W_f(\sigma) \leq W + 2\epsilon$. ◇

This yields a characterization of Riemann-integrability in which the 're-finement-limit' of 9.7.5 is replaced by a 'norm-limit':

9.7.10. Theorem. *For a bounded function* $f : [a, b] \to \mathbb{R}$, *the following conditions are equivalent*:[3]
   (a) f *is Riemann-integrable*;
   (b) $\lim_{N(\sigma) \to 0} W_f(\sigma) = 0$.

Proof. This is immediate from 9.7.9. ◇

Exercises

1. (i) Prove that if f is Riemann-integrable then so is $|f|$. {Hint: If $x, y \in [a_{\nu-1}, a_\nu]$ then

$$\big||f(x)| - |f(y)|\big| \leq |f(x) - f(y)| \leq \omega_\nu(f);$$

infer that $\omega_\nu(|f|) \leq \omega_\nu(f).$}
   (ii) Is the converse true?

2. Prove: If f is Riemann-integrable then so is f^2. {Hint: By Exercise 1, we can suppose that $f \geq 0$. Then

$$\omega_\nu(f^2) = [M_\nu(f) - m_\nu(f)][M_\nu(f) + m_\nu(f)];$$

infer that if $0 \leq f \leq K$ on $[a, b]$ then $W_{f^2}(\sigma) \leq 2K \cdot W_f(\sigma)$ for every subdivision σ.}

---

[3]This reformulation of Riemann-integrability is attributed to Riemann [T. Hawkins, op. cit, pp. 17, 40].

3. Prove that if f and g are Riemann-integrable on $[a, b]$ then so is $f + g$. {Hint: From the inequality

$$|(f + g)(x) - (f + g)(y)| \le |f(x) - f(y)| + |g(x) - g(y)|$$

deduce that $W_{f+g}(\sigma) \le W_f(\sigma) + W_g(\sigma)$ for every subdivision σ of $[a, b]$.}

It is harder to see that the integral of the sum is equal to the sum of the integrals; that's what Riemann sums are good for! (See the next section.)

4. Prove that if f and g are Riemann-integrable on $[a, b]$ then so is fg. {Hint: Exercises 2 and 3, Exercise 2 of §9.4, and the identity

$$fg = \frac{1}{4}[(f + g)^2 - (f - g)^2].$$

Alternatively, from the identity

$$(fg)(x) - (fg)(y) = [f(x) - f(y)]g(x) + f(y)[g(x) - g(y)]$$

deduce that $W_{fg}(\sigma) \le L \cdot W_f(\sigma) + K \cdot W_g(\sigma)$, where $|f| \le K$ and $|g| \le L$ on $[a, b]$.}

5. It was proved in 9.4.5 that if $f : [a, b] \to \mathbb{R}$ is continuous then it is Riemann-integrable. Give an alternative proof, using criterion (b) of 9.7.5, based on the uniform continuity of f. {Hint: Given any $\epsilon > 0$, choose $\delta > 0$ as in 6.6.1. If σ is any subdivision with $N(\sigma) < \delta$, argue that $W_f(\sigma) \le \epsilon(b - a)$.}

6. In the context of Riemann-Stieltjes integration with respect to an increasing function g (cf. §9.1, Exercise 5 and §9.4, Exercise 6), the formula for 'weighted oscillation' in Definition 9.7.1 becomes

$$W_f(\sigma) = S(\sigma) - s(\sigma) = \sum_{\nu=1}^{n} (M_\nu - m_\nu)[g(a_\nu) - g(a_{\nu-1})].$$

Modify the definition of 'norm' to take account of g, by defining

$$N_g(\sigma) = \max\{g(a_\nu) - g(a_{\nu-1}) : \nu = 1, \ldots, n\}$$

for a subdivision $\sigma = \{a = a_0 < a_1 < \cdots < a_n = b\}$. All results of this section then carry over to the Riemann-Stieltjes setting with $N(\sigma)$ replaced by $N_g(\sigma)$. In particular, the analogue of Theorem 9.7.10 may be stated as follows: A bounded function f is RS-integrable with respect to g if and only if $W_f(\sigma) \to 0$ as $N_g(\sigma) \to 0$.

7. In the spirit of Exercise 6, carry over the results of Exercises 1–5 to the Riemann-Stieltjes setting.[4]

{In the hint for Exercise 5, $N(\sigma)$ is defined as in 9.1.1, the inequality becomes $W_f(\sigma) \leq \epsilon[g(b) - g(a)]$, and the desired conclusion is obtained by citing the criterion (b) of the analogue of Theorem 9.7.5.}

9.8. The Integral as a Limit of Riemann Sums

The notations are those established at the beginning of the chapter (9.1.1–9.1.3).

9.8.1. *Definition.* If $\sigma = \{a = a_0 < a_1 < \cdots < a_n = b\}$ is a subdivision of $[a, b]$ and if, in each subinterval $[a_{\nu-1}, a_\nu]$, a point is selected, say $x_\nu \in [a_{\nu-1}, a_\nu]$ $(\nu = 1, \ldots, n)$, then the number

$$\sum_{\nu=1}^{n} f(x_\nu)e_\nu$$

is called a **Riemann sum** for f; it depends on f, on σ and on the selected points x_ν, as is indicated by writing

$$R_f(\sigma; x_1, \ldots, x_n) = \sum_{\nu=1}^{n} f(x_\nu)e_\nu$$

(verbally, a 'sum with selection').

Obviously $s_f(\sigma) \leq R_f(\sigma; x_1, \ldots, x_n) \leq S_f(\sigma)$, so it is not surprising that Riemann-integrability can be formulated in terms of such sums:

9.8.2. *Theorem. For a bounded function $f : [a, b] \to \mathbb{R}$, the following conditions are equivalent:*

(a) *f is Riemann-integrable;*

(b) *there exists a real number λ such that*

$$\lim_{N(\sigma)\to 0} R_f(\sigma; x_1, \ldots, x_n) = \lambda$$

in the following sense: for every $\epsilon > 0$ there exists a $\delta > 0$ such that

$$\left| R_f(\sigma; x_1, \ldots, x_n) - \lambda \right| \leq \epsilon$$

for every subdivision $\sigma = \{a = a_0 < a_1 < \cdots < a_n = b\}$ with $N(\sigma) \leq \delta$ and for every selection $x_\nu \in [a_{\nu-1}, a_\nu]$ $(\nu = 1, \ldots n)$.

---

[4]With a little extra effort in Exercise 3, one can obtain the formula $\int (f_1 + f_2)dg = \int f_1 dg + \int f_2 dg$, where f_1 and f_2 are RS-integrable with respect to g [cf. W. Rudin, *Principles of mathematical analysis*, Theorem 6.12, 3rd edn., McGraw-Hill, New York, 1976].

Necessarily

$$\lambda = \int_a^b f\,.$$

Proof. {Condition (b) is Riemann's original formulation of integrability.[1] }

(a) $\Rightarrow$ (b): Assuming f is Riemann-integrable, let

$$\lambda = \int_a^b f\,;$$

let us show that λ satisfies condition (b). For any subdivision $\sigma = \{a = a_0 < a_1 < \cdots < a_n = b\}$ and any selection $x_\nu \in [a_{\nu-1}, a_\nu]$, we have

$$s(\sigma) \leq R_f(\sigma; x_1, \ldots, x_n) \leq S(\sigma), \quad s(\sigma) \leq \lambda \leq S(\sigma)\,,$$

therefore

$$\left| R_f(\sigma; x_1, \ldots, x_n) - \lambda \right| \leq S(\sigma) - s(\sigma) = W_f(\sigma)\,.$$

Given any $\epsilon > 0$, choose $\delta > 0$ so that $W_f(\sigma) \leq \epsilon$ whenever $N(\sigma) \leq \delta$ (9.7.10); it is clear that δ meets the requirements of (b).

(b) $\Rightarrow$ (a): Assuming λ as in (b), we have to show that f is Riemann-integrable and that its integral is λ.

Given any $\epsilon > 0$, choose $\delta > 0$ as in (b). Let $\sigma = \{a = a_0 < a_1 < \cdots < a_n = b\}$ be a subdivision with $N(\sigma) \leq \delta$; to prove that f is Riemann-integrable, it suffices by 9.7.5 to show that $W_f(\sigma) \leq 2\epsilon$. For any selection $x_1, \ldots, x_n$ from the subintervals for σ, we know from (b) that

(*) $$\lambda - \epsilon \leq \sum_{\nu=1}^n f(x_\nu)e_\nu \leq \lambda + \epsilon\,.$$

For each $\nu = 1, \ldots, n$, choose a sequence (x_ν^k) in $[a_{\nu-1}, a_\nu]$ such that

$$f(x_\nu^k) \to M_\nu \text{ as } k \to \infty\,;$$

then

$$\sum_{\nu=1}^n f(x_\nu^k)e_\nu \to \sum_{\nu=1}^n M_\nu e_\nu = S(\sigma)\,,$$

so by (*) we have

(**) $$\lambda - \epsilon \leq S(\sigma) \leq \lambda + \epsilon\,.$$

---

[1] T. Hawkins, op. cit., p. 17.

Similarly,

$$\lambda - \epsilon \le s(\sigma) \le \lambda + \epsilon;$$

thus $S(\sigma)$ and $s(\sigma)$ both belong to the interval $[\lambda - \epsilon, \lambda + \epsilon]$, therefore

$$W_f(\sigma) = S(\sigma) - s(\sigma) \le 2\epsilon.$$

This proves that f is Riemann-integrable and since, by 9.7.8,

$$S(\sigma) \to \int_a^b f \quad \text{as} \quad N(\sigma) \to 0,$$

it is clear from (**) that

$$\lambda = \int_a^b f. \; \Diamond$$

9.8.3. Corollary. *If f and g are Riemann-integrable on $[a, b]$, then $f + g$ is also Riemann-integrable and*

$$\int_a^b (f + g) = \int_a^b f + \int_a^b g.$$

Proof.[2] Let

$$\lambda = \int_a^b f \quad \text{and} \quad \mu = \int_a^b g.$$

If $\sigma = \{a = a_0 < a_1 < \cdots < a_n = b\}$ is any subdivision of $[a, b]$ then, for every selection $x_\nu \in [a_{\nu-1}, a_\nu]$ ($\nu = 1, \dots n$), we have

$$R_{f+g}(\sigma; x_1, \dots, x_n) = \sum_{\nu=1}^{n} [f(x_\nu) + g(x_\nu)] e_\nu$$

$$= R_f(\sigma; x_1, \dots, x_n) + R_g(\sigma; x_1, \dots, x_n);$$

as $N(\sigma) \to 0$, the right member tends to $\lambda + \mu$, therefore

$$R_{f+g}(\sigma; x_1, \dots, x_n) \to \lambda + \mu$$

(all of this is easily said with ϵ's and δ's), and it follows that $f + g$ is Riemann-integrable, with integral $\lambda + \mu$ (9.8.2). $\Diamond$

For want of a similar formula for the integral of fg in terms of the integrals of f and g, Riemann sums have nothing to contribute to the proof that the integrability of f and g implies that of fg (see §9.7, Exercise 4).

---

[2]For a proof using upper and lower Darboux integrals, see H. Kestelman, *Modern theories of integration* [Oxford, 1937; reprinted Dover, 1960], p. 44.

9.8.4. *Remarks.* Condition (a) of 9.8.2 (equality of the lower and upper integrals) should perhaps be called 'Darboux-integrability', and condition (b), 'Riemann-integrability'. Theorem 9.8.2 says that the concepts are
equivalent, but there's still an important difference in the way the value
space $\mathbb{R}$ of the functions is treated: condition (a) relies on the order properties of $\mathbb{R}$ (via sups and infs), while condition (b) relies on the distance
properties (via absolute value). This opens the way for generalizing the
concept of integral in different directions: Darboux's way for value spaces
having order but not distance, Riemann's way for value spaces having distance but not order. This is a typical way in which an equivalence (P
$\Leftrightarrow$ Q) can serve as the point of departure for generalizations in different
directions.

9.8.5. *Postlude.* The last word on Riemann-integrability was uttered
by H. Lebesgue in the early 1900's: A bounded function $f : [a, b] \to \mathbb{R}$
is Riemann-integrable if and only if the set D of points of discontinuity
of f is "negligible". The term in quotes means that for every $\epsilon > 0$,
there exists a sequence of intervals (I_n) such that $D \subset \bigcup I_n$ and the sum
of the lengths of the intervals is $< \epsilon$. The proof (not difficult) is given in
the last chapter (11.4.1).

Exercises

1. If f and g are Riemann-integrable on $[a, b]$, then so are the functions
$$h(x) = \max\{f(x), g(x)\},$$
$$k(x) = \min\{f(x), g(x)\},$$
and
$$\int_a^b h + \int_a^b k = \int_a^b f + \int_a^b g.$$
{Hint: §2.9, Exercise 1 and §9.7, Exercises 1, 3.}

2. There is a 'Cauchy criterion' for the Riemann-integrability of a
bounded function f: in order that there exist a real number λ satisfying (b) of 9.8.2, it is necessary and sufficient that, for every $\epsilon > 0$, there
exist a $\delta > 0$ such that
$$\left| R_f(\sigma; x_1, \dots, x_n) - R_f(\tau; y_1, \dots, y_m) \right| \le \epsilon$$
for every pair of subdivisions
$$\sigma = \{a = a_0 < a_1 < \cdots < a_n = b\},$$
$$\tau = \{a = b_0 < b_1 < \cdots < b_m = b\},$$

with $N(\sigma) \leq \delta$ and $N(\tau) \leq \delta$, and for any selections $x_\nu \in [a_{\nu-1}, a_\nu]$, $y_\mu \in [b_{\mu-1}, b_\mu]$.

{Hint: Necessity is obvious (triangle inequality). Sufficiency: Let (σ_i) be a sequence of subdivisions of $[a, b]$ with $N(\sigma_i) \to 0$. For each i, make a selection for σ_i (for example, select the left endpoints of the subintervals for σ_i) and let

$$\lambda_i = R_f(\sigma_i; \text{ the selection for } \sigma_i).$$

Show that (λ_i) is a Cauchy sequence in $\mathbb{R}$, let λ be its limit, and show that λ meets the requirements of (b) of 9.8.2.}

3. If (x_n) is any sequence of points in $\mathbb{R}$, prove that the set $A = \{x_n : n \in \mathbb{P}\}$ has Lebesgue measure zero in the sense of 9.8.5.

4. Theorem 9.8.2 and its corollary are valid for the Riemann-Stieltjes integral $\int_a^b f \, dg$ (defined in §9.4, Exercise 6), with $N(\sigma)$ replaced by $N_g(\sigma)$ as in §9.7, Exercise 6.

{Hint: Since $e_\nu = g(a_\nu) - g(a_{\nu-1})$ in this setting (§9.1, Exercise 5), the sums-with-selection of 9.8.1 are given by $R_f(\sigma; x_1, \ldots, x_n) = \sum_{\nu=1}^{n} f(x_\nu) [g(a_\nu) - g(a_{\nu-1})]$.}

CHAPTER 10

Infinite Series

Infinite series are one of the grand themes of analysis. The most spectacular applications (Fourier series, power series, orthogonal series, ...) have to do with series of functions; this brief chapter touches only on the underlying fundamentals of series of constants.

If (s_n) is a convergent *sequence* of real numbers, the differences $s_n - s_{n-1}$ tell us something about the *speed* of convergence. Writing $a_n = s_n - s_{n-1}$ (with the convention $s_0 = 0$), the s_n can be recovered from the a_n via a telescoping sum $s_n = \sum_{k=1}^{n} a_k$, thus $s_n \to s$ means $\sum_{k=1}^{n} a_k \to s$ as $n \to \infty$ and it is natural to write

$$s = \sum_{k=1}^{\infty} a_k .$$

The theory of infinite series is the study of such 'infinite sums'.

10.1. Infinite Series: Convergence, Divergence

10.1.1. *Definition.* If (a_n) is a sequence of real numbers, the symbol

$$\sum_{n=1}^{\infty} a_n$$

(or $\sum_{1}^{\infty} a_n$, or simply $\sum a_n$) is called an **infinite series** (briefly, series); a_n is called the n'**th term** of the series. We write

$$s_n = \sum_{k=1}^{n} a_k ,$$

179

called the n'**th partial sum** of the series. One also writes

$$\sum_{n=1}^{\infty} a_n = a_1 + a_2 + a_3 + \ldots.$$

(This is an equality of symbols—they are equal *by definition*—not necessarily interpretable as an equality of numbers.)

10.1.2. Example. If $c \in \mathbb{R}$, $c \neq 1$, and if $a_n = c^{n-1}$ ($n = 1, 2, 3, \ldots$), then

$$s_n = 1 + c + c^2 + \ldots + c^{n-1} = \frac{1 - c^n}{1 - c}.$$

10.1.3. Definition. With notations as in 10.1.1, the series is said to be **convergent** if the sequence (s_n) of partial sums is convergent in $\mathbb{R}$; if $s_n \to s$ then s is called the **sum** of the series, we write

$$\sum_{n=1}^{\infty} a_n = s,$$

and we say that the series **converges** to s. A series that is not convergent is said to be **divergent**.

Thus, for a convergent series, the symbol $\sum_1^{\infty} a_n$ is assigned a numerical value; for a divergent series, we regard it just as a symbol.[1]

10.1.4. Example. If $|c| < 1$ then the series

$$\sum_{n=1}^{\infty} c^{n-1} = 1 + c + c^2 + c^3 + \ldots$$

is convergent, with sum $\frac{1}{1-c}$ (see the formula in 10.1.2). If $c = 1$ then $s_n = n$ and the series is divergent; if $c = -1$ then s_n alternates between 1 and 0, so the series is again divergent; and if $|c| > 1$ then (s_n) is unbounded, therefore divergent, so the series is again divergent. Briefly, the series is convergent when $|c| < 1$, divergent when $|c| \geq 1$. It is called a *geometric series* with *ratio* c (each term is obtained from its predecessor by multiplying by c).

10.1.5. Remarks. Let C be a condition that an infinite series $\sum a_n$ may or may not satisfy. If

$$C \text{ true} \Rightarrow \sum a_n \text{ convergent},$$

---

[1]Just the same, there are various devious ways of assigning numerical values to the symbol even for certain divergent series (the theory of 'summability'). Cf. Exercise 2.

then C is called a *sufficient condition* for convergence. If

$$\sum a_n \text{ convergent} \quad \Rightarrow \quad C \text{ true},$$

then C is called a *necessary condition* for convergence.[2] For example, the condition $\ll a_n = 0 \ (\forall n) \gg$ is sufficient for convergence, obviously not necessary; the condition $\ll (s_n)$ is bounded$\gg$ is a necessary condition for convergence, but it is not sufficient (10.1.4 with $c = -1$). An interesting necessary condition is $a_n \to 0$:

10.1.6. **Theorem.** *If* $\sum a_n$ *is convergent, then* $a_n \to 0$.

Proof. If (s_n) is the sequence of partial sums, say $s_n \to s$, then $a_{n+1} = s_{n+1} - s_n \to s - s = 0$, thus the sequence (a_n) is null. ◊

10.1.7. *Example.* The condition $a_n \to 0$ is not sufficient for the convergence of $\sum a_n$. For example, the series

$$1 + \frac{1}{2} + \frac{1}{3} + \ldots$$

(called the *harmonic series*) is divergent, although $a_n = 1/n \to 0$. For, the sequence (s_n) of partial sums satisfies

$$s_{2n} - s_n = \frac{1}{n+1} + \ldots + \frac{1}{n+n} = \sum_{k=1}^{n} \frac{1}{n+k} > n \cdot \frac{1}{n+n} = \frac{1}{2},$$

therefore is not convergent. {Proof #1: Cauchy's criterion. Proof #2: An easy 'telescoping' argument shows that $s_{2^k} - s_1 \geq k \cdot \frac{1}{2}$, whence (s_n) is unbounded.}

Cauchy's criterion for convergence of sequences (§3.6) yields a necessary and sufficient condition for the convergence of series (slightly reformulated in several ways):

10.1.8. **Theorem.** (Cauchy's criterion) *For an infinite series* $\sum a_n$, *the following conditions are equivalent:*
(a) $\sum a_n$ *is convergent;*
(b) *for every* $\epsilon > 0$, *there exists a positive integer* N *such that*

$$m \geq n \geq N \quad \Rightarrow \quad \left| \sum_{k=n}^{m} a_k \right| \leq \epsilon ;$$

(c) *for every* $\epsilon > 0$, *there exists a positive integer* N *such that*

$$n \geq N \quad \Rightarrow \quad \left| \sum_{k=n}^{n+p} a_k \right| \leq \epsilon \quad for \quad p = 0, 1, 2, 3, \ldots ;$$

---

[2]The language is obviously applicable to other situations: an implication P $\Rightarrow$ Q can be expressed by saying that Q is a necessary condition for P , and that P is a sufficient condition for Q .

(d) *for every* $\epsilon > 0$, *there exists a positive integer* N *such that*

$$n \geq N \quad \Rightarrow \quad \left| \sum_{k=n}^{n+p} a_k \right| \leq \epsilon \quad for \quad p = 1, 2, 3, \ldots .$$

Proof. If $m \geq n \geq N > 1$ and $p \in \mathbb{N}$, it is clear from the formulas

$$\sum_{k=n}^{m} a_k = s_m - s_{n-1}, \qquad \sum_{k=n}^{n+p} a_k = s_{n+p} - s_{n-1}$$

that the conditions (b)–(d) are reformulations of Cauchy's criterion for the convergence of the sequence (s_n). ◇

Exercises

1. Find the sum of the series $\sum_{n=1}^{\infty} 2/3^{n-1}$.

2. With notations as in 10.1.1, if $a_n = (-1)^{n+1}$ then the sequence (s_n) alternates between 1 and 0. Let t_n be the average of $s_1, \ldots, s_n$:

$$t_n = \frac{s_1 + s_2 + \ldots + s_n}{n}.$$

Then t_n alternates between $(n+1)/2n$ and $1/2$, so $t_n \to 1/2$. Why not be daring and write

$$1 + (-1) + 1 + (-1) + \ldots = \frac{1}{2}.$$

Much worse off is the series $1 + 1 + 1 + \ldots$ (what is t_n in this case?), and repetition of the averaging process doesn't improve matters.

3. Convergence is an 'ultimate' matter in the following sense: if $a_n = b_n$ ultimately, then $\sum a_n$ is convergent if and only if $\sum b_n$ is convergent. {Hint: If (s_n) and (t_n) are the respective sequences of partial sums, then $s_n - t_n$ is ultimately constant.}

4. (*Leibniz's alternating series test*) If $a_n \geq 0$ for n odd, $a_n \leq 0$ for n even, and $|a_n| \downarrow 0$, then the series $\sum a_n$ is convergent.
{Hint: If (s_n) is the sequence of partial sums, then the subsequence (s_{2n}) is increasing, (s_{2n-1}) is decreasing, and $s_2 \leq s_{2n} \leq s_{2n+1} \leq s_1$.}

10.2. Algebra of Convergence

10.2.1. **Theorem.** *If $\sum a_n$ and $\sum b_n$ are convergent, then so are $\sum (a_n + b_n)$ and $\sum (ca_n)$ $(c \in \mathbb{R})$, and*

$$\sum (a_n + b_n) = \sum a_n + \sum b_n , \quad \sum (ca_n) = c \sum a_n .$$

Proof. Let

$$s_n = \sum_{k=1}^{n} a_k , \quad t_n = \sum_{k=1}^{n} b_k$$

and suppose $s_n \to s$, $t_n \to t$. Then

$$\sum_{k=1}^{n} (a_k + b_k) = s_n + t_n \to s + t ,$$

and

$$\sum_{k=1}^{n} (ca_k) = cs_n \to cs$$

by 3.4.8. ◇

The second formula of 10.2.1 looks like an infinite 'distributive law'. There's also an 'associative law'; first, let's make precise the kind of 'associativity' that is involved:

10.2.2. *Definition.* Given an infinite series $\sum_1^\infty a_n$ and a strictly increasing sequence of positive integers

$$n_1 < n_2 < n_3 < \ldots ,$$

define

$$b_1 = a_1 + a_2 + \ldots + a_{n_1}$$
$$b_2 = a_{n_1+1} + \ldots + a_{n_2}$$
$$b_3 = a_{n_2+1} + \ldots + a_{n_3}$$
$$\ldots$$
$$b_k = a_{n_{k-1}+1} + \ldots + a_{n_k}$$

(with $n_0 = 0$). The series $\sum b_k$ is called a **regrouping** of the series $\sum a_n$. {The series $\sum b_k$ is obtained by adding the terms of $\sum a_n$ in finite 'packets'.}

The bad news: If $a_n = (-1)^{n+1}$ then the series

$$\sum_{n=1}^{\infty} a_n = 1 + (-1) + 1 + (-1) + \ldots$$

is divergent, but there is an obvious regrouping that is convergent (with sum 0). The good news:

10.2.3. Theorem. (Associative law) *If $\sum a_n$ is convergent, then every regrouping $\sum b_k$ of $\sum a_n$ is also convergent, with same sum.*

Proof. Let s_n be the n'th partial sum of $\sum a_n$, t_k the k'th partial sum of $\sum b_k$; with notations as in 10.2.2,

$$t_k = b_1 + \ldots + b_k = a_1 + \ldots + a_{n_k} = s_{n_k}$$

by the ordinary associative law for addition, thus (t_k) is a subsequence of the sequence (s_n). If $s_n \to s$ then also $t_k \to s$ (3.5.5). $\Diamond$

There is also a restricted version of 'commutativity' (10.4.3) and a kind of multiplication for series (10.4.5), but both require an extra concept ('absolute convergence').

Exercises

1. If the sequence of partial sums of an infinite series is bounded, then the series has a convergent regrouping. {Hint: Weierstrass-Bolzano.}

2. Suppose the series $\sum b_n$ is obtained from a series $\sum a_n$ by inserting a finite number of zero terms between a_n and a_{n+1}; thus, there is a subsequence (b_{n_k}) of (b_n) such that $b_{n_k} = a_k$ ($k = 1, 2, 3, \ldots$) and $b_n = 0$ when n is not equal to any n_k. Prove that $\sum a_n$ converges if and only if $\sum b_n$ converges, in which case $\sum a_n = \sum b_n$.
{Hint: If $s_k = a_1 + \ldots + a_k$ and $t_n = b_1 + \ldots + b_n$, then $s_k = t_{n_k}$ for all k, and $t_n = s_k$ for $n_k \le n < n_{k+1}$.}

10.3. Positive-Term Series

10.3.1. *Definition.* One calls $\sum a_n$ a **positive-term series** if $a_n \ge 0$ for all n.
The 'fundamental theorem' of such series:

10.3.2. Theorem. *A positive-term series $\sum a_n$ is convergent if and only if its sequence (s_n) of partial sums is bounded; for such a series,*

$$\sum_{n=1}^{\infty} a_n = \sup s_n,$$

thus $s_n \uparrow \sum_1^{\infty} a_n$.

Proof. The sequence (s_n) is increasing (because $a_n \geq 0$), so it is convergent if and only if it is bounded (3.4.3), in which case $\lim s_n = \sup s_n$. ◊

10.3.3. *Convention.* For a positive-term series, we write $\sum_1^\infty a_n < +\infty$ if the series converges, $\sum_1^\infty a_n = +\infty$ if it diverges. {In either case, $\sum_1^\infty a_n = \sup s_n$ in the ordered set $\mathbb{R} \cup \{-\infty, +\infty\}$ of 'extended real numbers' (4.1.1).}

10.3.4. *Example.* $\sum_{n=1}^\infty 1/n^2 < +\infty$.
{Proof:[1] Writing s_n for the n'th partial sum, let's show that $s_n < 2$ for all n. This is trivial for $n = 1$, and

$$s_{n+1} = \sum_{k=1}^{n+1} \frac{1}{k^2} = 1 + \sum_{k=2}^{n+1} \frac{1}{k^2}$$

$$< 1 + \sum_{k=2}^{n+1} \frac{1}{(k-1)k}$$

$$= 1 + \sum_{k=2}^{n+1} \left[\frac{1}{k-1} - \frac{1}{k} \right]$$

$$= 1 + \sum_{k=1}^{n} \frac{1}{k} - \sum_{k=2}^{n+1} \frac{1}{k}$$

$$= 1 + \frac{1}{1} - \frac{1}{n+1} < 2$$

(telescoping sum).}

A partial sum selects *the first n terms*; a 'finite subsum' selects any finite set of terms:

10.3.5. *Definition.* Let $\sum a_n$ be any infinite series. For every finite set of indices $F \subset \mathbb{P}$, write

$$s_F = \sum_{k \in F} a_k \, ;$$

such an s_F is called a **finite subsum** of the series $\sum a_n$. {Convention: $s_\varnothing = 0$.} Note that if $F_n = \{1, 2, \ldots, n\}$ then $s_{F_n} = s_n$ (the n'th partial sum).

For a positive-term series, the supremum of *all* the finite subsums is equal to the supremum of *some* of them:

10.3.6. *Theorem.* *If $\sum a_n$ is a positive-term series and (s_n) is its sequence of partial sums, then*

$$\sup s_n = \sup\{s_F : \, F \subset \mathbb{P}, \; F \; finite \, \}.$$

---

[1] E. Landau, *Differential and integral calculus*, p. 38.

Proof. Recall that $+\infty$ is allowed as a supremum (10.3.3). Note also that if F and G are finite subsets of $\mathbb{P}$ with $F \subset G$ then $s_F \le s_G$ (because $a_n \ge 0$). Write $s = \sup s_n$, t for the supremum of all the s_F. Since

$$s_n = s_{F_n} \le t \quad \text{for all} \quad n,$$

we have $s \le t$. On the other hand, if F is any finite subset of $\mathbb{P}$ then $F \subset F_n$ for some n, so

$$s_F \le s_{F_n} = s_n \le s;$$

varying F, $t \le s$. $\Diamond$

10.3.7. **Corollary.** *For a positive-term series* $\sum a_n$, *the following conditions are equivalent:*
(a) $\sum a_n < +\infty$;
(b) *the set of finite subsums* s_F *is bounded.*
For such a series,

$$\sum a_n = \sup\{s_F : F \subset \mathbb{P}, F \text{ finite}\}.$$

Proof. This is immediate from 10.3.2 and 10.3.6. $\Diamond$

10.3.8. **Corollary.** (Commutative law) *Given a positive-term series* $\sum a_n$ *and a permutation* σ *of the set* $\mathbb{P}$ *of positive integers (that is, a bijection* $\sigma : \mathbb{P} \to \mathbb{P}$), *define* $b_n = a_{\sigma(n)}$ *and consider the series* $\sum b_n$ *(called a* **rearrangement** *of the original series). Then*

$$\sum a_n = \sum b_n.$$

In particular, every rearrangement of a convergent positive-term series is convergent, with same sum.

Proof. If F is a finite subset of $\mathbb{P}$, write

$$s_F = \sum_{n \in F} a_n, \quad t_F = \sum_{n \in F} b_n.$$

Then

$$t_F = \sum_{n \in F} a_{\sigma(n)} = \sum_{k \in \sigma(F)} a_k = s_{\sigma(F)},$$

so the set of finite subsums for the two series is the same, and the corollary is immediate from 10.3.7. $\Diamond$

For series whose terms are not all positive, a rearrangement of a convergent series may diverge (Exercise 1).

Exercises

1. The convergent series

$$\frac{1}{2} - \frac{1}{2} + \frac{1}{3} - \frac{1}{3} + \frac{1}{4} - \frac{1}{4} + \cdots$$

has a divergent rearrangement

$$\frac{1}{2} - \frac{1}{2} + \frac{1}{3} + \frac{1}{4} - \frac{1}{3} + \frac{1}{5} + \frac{1}{6} + \frac{1}{7} + \frac{1}{8} - \frac{1}{4} + \cdots.$$

{In the second series, the appearance of the positive terms is accelerated so as to create packets of successive terms with sum $> 1/2$ (cf. 10.1.7).}

2. (*Comparison test*) Assume $0 \leq a_n \leq b_n$ for all n (or just ultimately). Then

$$\sum b_n \text{ convergent} \implies \sum a_n \text{ convergent},$$

in other words,

$$\sum a_n \text{ divergent} \implies \sum b_n \text{ divergent}.$$

3. (i) If (d_n) is a sequence of nonnegative integers such that $0 \leq d_n \leq 9$ for all n, then the series

$$\sum_{n=1}^{\infty} \frac{d_n}{10^n}$$

is convergent; its sum is denoted $.d_1 d_2 d_3 \ldots.$ {Hint: $0 \leq d_n/10^n \leq 9/10^n$.}

(ii) With notations as in (i), $0 \leq .d_1 d_2 d_3 \ldots \leq 1$. If $x = .d_1 d_2 d_3 \ldots$, $d_n = 9$ ultimately and $0 < x < 1$, then x has a representation in which $d_n = 0$ ultimately.

(iii) If $x = .d_1 d_2 d_3 \ldots$ and $d_n < 9$ frequently, then $d_1 = [10x]$, $d_2 = [10^2 x] - d_1 \cdot 10^1$, $d_3 = [10^3 x] - d_1 \cdot 10^2 - d_2 \cdot 10^1$, etc.}

(iv) Every $x \in [0,1]$ can be written $x = .d_1 d_2 d_3 \ldots$ for a suitable sequence (d_n).

{Hint: We can suppose that $0 < x < 1$. Let d_1 be the largest nonnegative integer such that $d_1 10^{-1} \leq x$ and write $x_1 = d_1 10^{-1}$. Assuming $d_1, \ldots d_{n-1}$ already defined and $x_{n-1} = d_1 10^{-1} + \ldots + d_{n-1} 10^{-(n-1)}$, let d_n be the largest nonnegative integer such that $x_{n-1} + d_n 10^{-n} \leq x$ and write $x_n = x_{n-1} + d_n 10^{-n}$. Then $x_{n-1} + (d_n + 1) 10^{-n} > x$, whence $x - x_n < 10^{-n}$. Necessarily $d_n \leq 9$; for, if $d_n \geq 10$ then $d_n 10^{-n} \geq 10^{-(n-1)}$ leads to

$$x \geq x_n \geq x_{n-2} + (d_{n-1} + 1)10^{-(n-1)},$$

contrary to the maximality of d_{n-1}.}

4. Describe the set of finite subsums of the series $\sum_1^\infty (-1)^n$.

5. Deduce another proof of $\sum 1/n^2 < +\infty$ from the observation that

$$\int_1^n \frac{1}{x^2} dx = 1 - \frac{1}{n} < 1$$

for every positive integer n. {It can be shown[2] that the *value* of the series is $\pi^2/6$.}

6. Deduce that $\sum 1/\sqrt{n} = +\infty$ by adapting the trick in Exercise 5. Then find a simpler proof (cf. Exercise 2).

7. The series $\sum 1/\ln(1+n)$ is divergent. {Hint: 9.5.3, (iv).}

8. If $\sum a_n$ is a convergent positive-term series whose sequence of terms is decreasing (hence $a_n \downarrow 0$), then $na_n \to 0$.
{Hint: The sequence (s_n) of partial sums is increasing and convergent. If $m > n$ then

$$s_m - s_n = a_{n+1} + \ldots + a_m \geq (m-n)a_m,$$

whence $ma_m \leq (s_m - s_n) + na_m$. Given $\epsilon > 0$, the problem is to show that $ma_m \leq \epsilon$ ultimately; think of n as a useful parameter.}

9. Let $\alpha \in \mathbb{R}$. The series $\sum 1/n^\alpha$ is (i) divergent if $\alpha \leq 1$, and (ii) convergent if $\alpha > 1$. (For the definition of n^α, see 9.5.13.)
{Hint: (i) Exercise 8. (ii) Use the method of Exercise 5; to calculate the integral in question, cite §9.5, Exercise 2.}

10. (*Comparison test of the second kind*) Suppose $a_n > 0$, $b_n > 0$ and $a_{n+1}/a_n \leq b_{n+1}/b_n$ for all n. Prove that if $\sum b_n$ is convergent, then so is $\sum a_n$.
{Hint: Multiply the given inequalities for $k = 1, \ldots, n$ to get $a_{n+1}/a_1 \leq b_{n+1}/b_1$ ('telescoping product').}

11. Let $\sum a_n$ be a series with $a_n > 0$ for all n.
(i) If $a_{n+1}/a_n \geq 1$ ultimately, then $\sum a_n$ is divergent.
(ii) If a_{n+1}/a_n is bounded and $\limsup(a_{n+1}/a_n) < 1$, then $\sum a_n$ is convergent (*Cauchy's ratio test*).
(iii) Give an example where $\sum a_n$ converges, a_{n+1}/a_n is bounded, but $\limsup(a_{n+1}/a_n) > 1$.
(iv) Calculate $\limsup(a_{n+1}/a_n)$ for the convergent series $\sum 1/n^2$ and for the divergent series $\sum 1/n$.

---

[2]K. Knopp, *Theory and application of infinite series* [2nd. English edition, Blackie, London, 1951], p. 237.

(v) Give an example where $\sum a_n$ is convergent but a_{n+1}/a_n is unbounded.

{Hints: (ii) If $\limsup(a_{n+1}/a_n) < r < 1$ then $a_{n+1}/a_n < r$ ultimately (§3.7, Exercise 7).

(iii) Assuming $b_n > 0$, $\sum b_n$ convergent and b_{n+1}/b_n bounded (for example, $b_n = c^n$, where $0 < c < 1$), let $a_{2n-1} = b_n$ and $a_{2n} = 2b_n$ for all n.

(v) Assuming $b_n > 0$ and $\sum n b_n$ convergent (for example, $b_n = 1/n^3$), let $a_{2n-1} = b_n$ and $a_{2n} = n b_n$ for all n.}

12. (*Cauchy's root test*) Let $\sum a_n$ be a positive-term series.

(i) If $(a_n)^{1/n} > 1$ frequently, then $\sum a_n$ is divergent.

(ii) If $(a_n)^{1/n}$ is bounded and $\limsup(a_n)^{1/n} < 1$, then $\sum a_n$ is convergent.

(iii) If $(a_n)^{1/n}$ is bounded and $\limsup(a_n)^{1/n} > 1$, then $\sum a_n$ is divergent.

(iv) Calculate $\limsup(a_n)^{1/n}$ for the convergent series $\sum 1/n^2$ and for the divergent series $\sum 1/n$.

{Hints: (i) If $a_n \to 0$ then $a_n \leq 1$ ultimately. (ii) If $\limsup(a_n)^{1/n} < r < 1$ then $(a_n)^{1/n} < r$ ultimately (§3.7, Exercise 7). (iii) If $\limsup (a_n)^{1/n} > r > 1$ then $(a_n)^{1/n} > r$ frequently. (iv) §9.5, Exercise 5.}

13. For every real number $c \geq 0$, the series $\sum_0^\infty c^n/n!$ is convergent.

{Hint: Proof #1: Let $a_n = c^n/n!$, $b_n = 1/2^n$ and apply Exercise 10. Proof #2: Exercise 11.}

14. If $a_n = n^3/3^n$ then the series $\sum a_n$ is convergent.

10.4. Absolute Convergence

10.4.1. *Definition.* A series $\sum a_n$ is said to be **absolutely convergent** if $\sum |a_n| < +\infty$ (cf. 10.3.3).

10.4.2. *Theorem.* (Triangle inequality) *If the series $\sum a_n$ is absolutely convergent then it is convergent, and $\left| \sum a_n \right| \leq \sum |a_n|$.*

Proof. Write

$$s_n = \sum_1^n a_k, \quad t_n = \sum_1^n |a_k|, \quad t = \sup t_n.$$

We are assuming that $t < +\infty$; we have to show that (s_n) is convergent and that $|\lim s_n| \leq t$. If $m \geq n$ then

$$\left| \sum_{k=n}^m a_k \right| \leq \sum_{k=n}^m |a_k|$$

by the ordinary triangle inequality; it is then clear from Cauchy's criterion (10.1.8) that the convergence of $\sum |a_n|$ implies that of $\sum a_n$. Writing $s = \lim s_n$, we have $|s_n| \to |s|$; since $|s_n| \leq t_n \leq t$ for all n, we conclude that $|s| \leq t$ (3.4.8). ◊

10.4.3. Theorem. (Commutative law) *If* $\sum a_n$ *is absolutely convergent, then every rearrangement is convergent (in fact, absolutely), with same sum.*

Proof. Assuming $\sum |a_n| < +\infty$ and $\sigma : \mathbb{P} \to \mathbb{P}$ bijective, let $b_n = a_{\sigma(n)}$. We know from 10.3.8 that $\sum |b_n|$ is convergent, therefore $\sum b_n$ is convergent by the preceding theorem; our problem is to show that $\sum a_n = \sum b_n$. Let

$$s_n = \sum_1^n a_k, \quad t_n = \sum_1^n b_k.$$

Say $s_n \to s$, $t_n \to t$; we must show that $s = t$, in other words, $s_n - t_n \to 0$.

Let $\epsilon > 0$. We seek an index p such that

$$n \geq p \;\Rightarrow\; |s_n - t_n| \leq \epsilon.$$

Since $\sum |a_n|$ is convergent, by 10.1.8 there is an index N such that

$$(*) \qquad\qquad m \geq n \geq N \;\Rightarrow\; \sum_{k=n}^m |a_k| \leq \epsilon.$$

Let $p_i = \sigma^{-1}(i)$ for $i = 1, \ldots N$ and let $p = \max\{p_1, \ldots p_N\}$; since the p_i are distinct, $p \geq N$. Also, since $p \geq p_i$ for $i = 1, \ldots, N$, we have

$$\{\sigma(1), \sigma(2), \ldots, \sigma(p)\} \supset \{\sigma(p_1), \ldots, \sigma(p_N)\} = \{1, 2, \ldots, N\},$$

therefore the list

$$b_1 = a_{\sigma(1)}, \; \ldots, \; b_p = a_{\sigma(p)}$$

includes the list $a_1, \ldots, a_N$.

Let $n \geq p$ and consider

$$(**) \qquad\qquad t_n - s_n = \sum_1^n b_k - \sum_1^n a_k.$$

Since $n \geq p$, the list $b_1, \ldots, b_n$ includes the list $b_1, \ldots, b_p$, hence also the list $a_1, \ldots, a_N$; and $n \geq p \geq N$, so the list $a_1, \ldots, a_n$ also includes $a_1, \ldots, a_N$. Thus $a_1, \ldots, a_N$ appear in both sums of $(**)$, hence cancel out; so for $n \geq p$, $t_n - s_n$ is a sum of terms $\pm a_k$ with $k > N$, consequently $|t_n - s_n| \leq \epsilon$ by $(*)$. ◊

10.4.4. *Notation.* So far, the terms of a series have been 'indexed' by the positive integers (a_n is the term of 'index' n). Often it is useful to start off the indices at 0; for example, we write

$$\sum_{n=1}^{\infty} c^{n-1} = \sum_{n=0}^{\infty} c^n .$$

More generally, we write

$$\sum_{n=0}^{\infty} a_n$$

to indicate the series $\sum_{1}^{\infty} b_n$ with $b_n = a_{n-1}$.

This notation is useful in formulating the following theorem about 'multiplying' series (a nice application of absolute convergence):

10.4.5. Theorem. (Mertens' theorem[1]) *Suppose the series*

$$\sum_{n=0}^{\infty} |a_n| \quad and \quad \sum_{n=0}^{\infty} b_n$$

are convergent. Define

$$c_0 = a_0 b_0$$
$$c_1 = a_0 b_1 + a_1 b_0$$
$$c_2 = a_0 b_2 + a_1 b_1 + a_2 b_0$$

and, in general, for every $n \in \mathbb{N}$,

$$c_n = \sum_{i+j=n} a_i b_j = a_0 b_n + a_1 b_{n-1} + \ldots + a_{n-1} b_1 + a_n b_0 .$$

Then the series $\sum_{0}^{\infty} c_n$ *is also convergent, and*

$$\sum_{n=0}^{\infty} c_n = \left(\sum_{n=0}^{\infty} a_n \right) \left(\sum_{n=0}^{\infty} b_n \right) .$$

Proof. Write

$$A_n = \sum_{k=0}^{n} a_k , \quad A = \sum_{n=0}^{\infty} a_n$$

$$B_n = \sum_{k=0}^{n} b_k , \quad B = \sum_{n=0}^{\infty} b_n$$

$$C_n = \sum_{k=0}^{n} c_k .$$

---

[1]K. Knopp, *Theory and application of infinite series*, p. 321; E. Landau, *Differential and integral calculus*, p. 163, Th. 220.

We know that $A_n \to A$ and $B_n \to B$; our problem is to show that $C_n \to AB$. We have

$$
\begin{aligned}
C_n &= c_0 + c_1 + c_2 + \ldots + c_n \\
&= (a_0 b_0) + (a_0 b_1 + a_1 b_0) + \ldots + (a_0 b_n + a_1 b_{n-1} + \ldots + a_{n-1} b_1 + a_n b_0) \\
&= a_0(b_0 + b_1 + \ldots + b_n) \\
&\quad + a_1(b_0 + b_1 + \ldots + b_{n-1}) \\
&\quad + \ldots \\
&\quad + a_{n-1}(b_0 + b_1) \\
&\quad + a_n b_0 \\
&= a_0 B_n + a_1 B_{n-1} + a_2 B_{n-2} + \ldots + a_{n-1} B_1 + a_n B_0 .
\end{aligned}
$$

Write $\beta_n = B_n - B$; thus $B_n = B + \beta_n$, where $\beta_n \to 0$, and

$$
\begin{aligned}
C_n &= a_0(B + \beta_n) + a_1(B + \beta_{n-1}) + \ldots + a_n(B + \beta_0) \\
&= (a_0 + a_1 + \ldots + a_n)B + (a_0 \beta_n + a_1 \beta_{n-1} + \ldots + a_n \beta_0) \\
&= A_n B + \gamma_n ,
\end{aligned}
$$

where

$$
(*) \qquad\qquad \gamma_n = a_0 \beta_n + a_1 \beta_{n-1} + \ldots + a_n \beta_0 .
$$

Since $A_n B \to AB$, it will suffice to show that $\gamma_n \to 0$.

We are thus reduced to proving the following proposition: If $\sum_0^\infty |a_n| < +\infty$ and $\beta_n \to 0$, and if γ_n is defined by (*), then $\gamma_n \to 0$. {The indices for (β_n) and (γ_n) start at 0.} Let $a = \sum_0^\infty |a_n|$; we can suppose $a > 0$. Given any $\epsilon > 0$, we have to show that $|\gamma_n| \leq \epsilon$ ultimately. Choose an index N such that

$$
k \geq N \implies |\beta_k| \leq \epsilon/2a .
$$

If $n > N$ then

$$
\gamma_n = (\beta_0 a_n + \beta_1 a_{n-1} + \ldots + \beta_{N-1} a_{n-(N-1)}) + (\beta_N a_{n-N} + \ldots + \beta_n a_0) .
$$

The first expression in parentheses has fixed length N and (since $a_n \to 0$) it is a linear combination of N null sequences (with coefficients $\beta_0, \ldots, \beta_{N-1}$), so its absolute value can be made $\leq \epsilon/2$ by taking n sufficiently large; by the choice of N,

$$
|\beta_N a_{n-N} + \ldots + \beta_n a_0| \leq (\epsilon/2a)(|a_{n-N}| + \ldots + |a_0|) \leq (\epsilon/2a)a = \epsilon/2 ,
$$

consequently (for n sufficiently large), $|\gamma_n| \leq \epsilon/2 + \epsilon/2$. $\diamond$

10.4.6. The series $\sum c_n$ of the preceding theorem is called the *Cauchy product* of the series $\sum a_n$ and $\sum b_n$. The conclusion of the theorem is a sort of 'distributive law' (imagine multiplying the series by 'doing what comes naturally').

Exercises

1. The converse of 10.4.3 is true: If every rearrangement of $\sum a_n$ is convergent, then $\sum |a_n|$ is convergent.[2]

2. In the notations of 10.4.5, if both $\sum a_n$ and $\sum b_n$ are absolutely convergent, then so is $\sum c_n$.

3. In the notations of 10.4.5, if both $\sum a_n$ and $\sum b_n$ are convergent but neither is absolutely convergent, then $\sum c_n$ may not be convergent. {Hint: Look at the Cauchy product of the convergent series

$$1 - \frac{1}{\sqrt{2}} + \frac{1}{\sqrt{3}} - \frac{1}{\sqrt{4}} + \ldots$$

with itself.}

4. True or false (explain): If the series

$$\sum_{n=0}^{\infty} a_n \quad \text{and} \quad \sum_{n=0}^{\infty} |b_n|$$

are convergent, then so is the series $\sum c_n$ of 10.4.5.

5. If $\sum a_n$ converges but not absolutely (such series are called *conditionally convergent*) then, for every real number b, some rearrangement of $\sum a_n$ converges to b. {The proof is not easy.[2]}

6. For every real number x, the series $\sum_0^{\infty} x^n/n!$ is absolutely convergent.

7. If $\sum a_n$ is absolutely convergent and (b_n) is a bounded sequence, then $\sum b_n a_n$ is absolutely convergent.

8. (i) If $a_n = (-1)^n/\ln(n+1)$, then the series $\sum a_n$ is conditionally convergent.
(ii) If $a_n = [(\sin n)/n]^n$ then the series $\sum a_n$ is absolutely convergent.

---

[2]E. Landau, op. cit., p. 158, Th. 217.

CHAPTER 11

Beyond the Riemann Integral

To motivate the chapter, let us recapitulate the 'Fundamental theorem of calculus' for continuous functions on a closed interval (cf. 9.4.6).

Let $f : [a, b] \to \mathbb{R}$ be a continuous function. The key to the fundamental theorem is the following consequence of the mean value theorem: If F and G are continuous real-valued functions on $[a, b]$ such that $F'(x) = f(x)$ and $G'(x) = f(x)$ for all points x in the open interval (a, b), then $F - G$ is a continuous function on $[a, b]$ such that $(F - G)'(x) = 0$ on (a, b), consequently $F - G$ is a constant function (8.5.4); in particular, the equality

$$(F - G)(a) = (F - G)(b)$$

means that $F(b) - F(a) = G(b) - G(a)$, so that $F(b) - F(a)$ depends only on f and not on the particular continuous function F satisfying $F' = f$ on (a, b). Free of charge, F has a one-sided derivative at the endpoints a and b, equal to $f(a)$ and $f(b)$, respectively (9.4.6).

The functions F that appear in this way, associated with all possible continuous functions f, are precisely the continuous functions F on $[a, b]$ having a finite derivative at every point of the interval (one-sided at the endpoints) and such that the derivative function F' is itself continuous on $[a, b]$; such functions F are conveniently described as the **continuously differentiable** functions on $[a, b]$.

The role of the Riemann integral in the foregoing is to show that for every continuous function f on $[a, b]$, a function F such that $F' = f$ on $[a, b]$ **exists** (9.3.4 and 9.3.5), namely

$$F(x) = \int_a^x f \qquad (a \le x \le b).$$

194

At the same time, the definition of the Riemann integral by means of upper and lower sums establishes the number $F(b) - F(a)$ as a plausible definition for the 'signed area' under the graph of f.

In the 'uniqueness' argument given above, the presence of the function f is superfluous. What counts is that any two continuous functions F and G on $[a, b]$ having the same derivative on (a, b) differ only by a constant, so that $F(b) - F(a) = G(b) - G(a)$; this requires only the mean value theorem (without reference to a function f or its Riemann integral).

This suggests extending the elementary integral in the following way. Call a function $f : [a, b] \to \mathbb{R}$ «integrable» if there exists a continuous function $F : [a, b] \to \mathbb{R}$ such that $F'(x) = f(x)$ for every x in (a, b); by the mean value theorem, one can then define without ambiguity the «integral» of f to be the number $F(b) - F(a)$, and, if we like, we can write expressions such as $\int_a^b f$ for this number. In particular, continuous functions f are «integrable» in this sense, with «integral» equal to the integral provided by the Riemann theory. A function that is «integrable» in this sense need not be continuous (Exercise 1), thus the class of «integrable» functions is larger than the class of continuous functions. An «integrable» function need not be Riemann-integrable, or even bounded (Exercise 2), so some non-Riemann-integrable functions acquire an «integral»; however, this good news masks a lot of bad news: many (in a sense, most) Riemann-integrable functions fail to be «integrable».[1] Thus, we now have two 'little theories' going off in separate directions instead of a single unified theory: the class of Riemann-integrable functions and the class of «integrable» functions both extend the class of continuous functions, but neither of these extensions is contained in the other.

The problem is that in the definition of «integrable» proposed above, the requirement that $F' = f$ on *all* of (a, b) is too restrictive and the requirement that F be *continuous* on $[a, b]$ is not restrictive enough. By relaxing 'all' to 'almost all' (in an appropriate sense) and by requiring F to be 'absolutely continuous' (in an appropriate sense), we can arrive at a class of functions f, much larger than the class of Riemann-integrable functions, for which a useful concept of integral can be defined. These are the *Lebesgue-integrable* functions (to be defined in §11.5). All of the cunning resides in the definition of 'almost all'; so to speak, the exceptional set of points x where F fails to have a derivative, or where it has a derivative different from $f(x)$, must be 'negligible' in an appropriate sense. The concept of 'absolute continuity' can be formulated in terms of 'negligible sets' as well.[2] Thus, the concept of 'negligible set' is the foundation of the entire Lebesgue theory; that's the place to begin.

---

[1] For an example, see B.R. Gelbaum and J.M.H. Olmsted, *Counterexamples in analysis* [Holden-Day, San Francisco, 1964], p. 43, Example 3.

[2] See §11.2, Exercise 4.

11.1. Negligible Sets

A set of real numbers is said to be negligible if it can be covered by a sequence of intervals whose total length is as small as we like. Formally:

11.1.1. *Definition.* A subset N of $\mathbb{R}$ is said to be **negligible** if, given any $\epsilon > 0$, there exists a sequence (I_n) of bounded intervals such that

$$N \subset \bigcup_{n=1}^{\infty} I_n \quad \text{and} \quad \sum_{n=1}^{\infty} \ell(I_n) \le \epsilon,$$

where $\ell(I) = b - a$ for an interval I with endpoints $a \le b$.

11.1.2. *Remarks.* If an interval does not contain one of its endpoints, the length is not changed by adjoining it; thus one can require the intervals in the above definition to be closed. On the other hand if, in the above definition, I_n has endpoints $a_n \le b_n$ and if one replaces I_n by the open interval

$$(a_n - \epsilon/2^{n+1}, a_n + \epsilon/2^{n+1}),$$

the total length is then $\le 2\epsilon$; thus one can require the intervals I_n to be open.

Putting these thoughts together, it is clear that the intervals in the definition can be required (permitted) to be any one of the four kinds (open, closed, semiclosed on the left, semiclosed on the right) or any mixture of the four kinds.

11.1.3. *Example.* Any subset N of $\mathbb{R}$ whose elements are the terms of a sequence is negligible. (Such sets are called *countable.*)

For, if $N = \{a_n : n = 1, 2, 3, \ldots\}$ then the (degenerate) intervals $I_n = [a_n, a_n] = \{a_n\}$ cover N and have total length 0.

For an example of a negligible set that is not countable, see Exercise 3.

11.1.4. *Remarks.* Negligible sets already play a decisive role in the theory of the Riemann integral. H. Lebesgue gave the following criterion for Riemann-integrability: *A bounded function $f : [a, b] \to \mathbb{R}$ is Riemann-integrable if and only if its set of discontinuities is negligible*, that is, the set

$$D_f = \{x \in [a, b] : f \text{ is not continuous at } x\}$$

is negligible. The proof (not difficult) is given in §11.4.

Here are some useful facts of the type 'new negligible sets from old':

11.1.5. *Proposition.* (1) *If N is a negligible set then every subset M of N is also negligible.*

(2) *If M and N are negligible sets, then so is their union $M \cup N$. More generally,*

(3) *If* (N_k) *is a sequence of negligible sets, then their union* $N = \bigcup N_k$
is also negligible.

Proof. (1) If $N \subset \bigcup I_k$ then also $M \subset \bigcup I_k$.

(2) Let $\epsilon > 0$. Since M is negligible, there exists a sequence (I_n) of intervals such that

$$M \subset \bigcup I_n \quad \text{and} \quad \sum \ell(I_n) \leq \epsilon/2.$$

Similarly, there exists a sequence (J_n) of intervals such that

$$N \subset \bigcup J_n \quad \text{and} \quad \sum \ell(J_n) \leq \epsilon/2.$$

Then $M \cup N$ is contained in the union of the sequence of intervals

$$I_1, J_1, I_2, J_2, I_3, J_3, \ldots,$$

the sum of whose lengths is $\leq \epsilon/2 + \epsilon/2 = \epsilon$.

(3) Given any $\epsilon > 0$, for each index k choose a sequence $I_{k1}, I_{k2}, I_{k3}, \ldots$ of intervals such that

$$N_k \subset \bigcup_{n=1}^{\infty} I_{kn} \quad \text{and} \quad \sum_{n=1}^{\infty} \ell(I_{kn}) \leq \epsilon/2^k;$$

then $N \subset \bigcup_{k,n} I_{kn}$ and the intervals I_{kn} $(k, n = 1, 2, 3, \ldots)$ can be arranged in a sequence (J_i). (For example, list those I_{kn} with $k+n = 2$, then those with $k + n = 3$, and so on.) Then

$$\sum_{i=1}^{\infty} \ell(J_i) \leq \sum_{k=1}^{\infty} \left(\sum_{n=1}^{\infty} \ell(I_{kn}) \right) \leq \sum_{k=1}^{\infty} \epsilon/2^k = \epsilon.$$

{To prove the first $\leq$ relation (actually, equality holds, but we can get by with $\leq$ here), it suffices to show that each partial sum $\sum_{i=1}^{r} \ell(J_i)$ of the series on the left side is $\leq$ the 'iterated sum' on the right side; this is easily seen by grouping together those intervals among $J_1, \ldots, J_r$ that come from the covering (I_{1n}) of N_1, then those that come from (I_{2n}), and so on until we have exhausted the list $J_1, \ldots, J_r$.} ◇

It follows that if a finite number (or even a sequence) of points are adjoined to or deleted from a negligible set, the resulting set is also negligible; in particular, in discussing negligible subsets of an interval $[a, b]$, we need not worry whether or not the endpoints of the interval belong to the subset.

The following 'obvious' property of negligible sets proves to be very useful in §11.3:

11.1.6. Proposition. *The interior of a negligible set is empty.*

Proof. Assuming N negligible and $[a, b] \subset N$ $(a \leq b)$, we need only show that $a = b$. By (1) of the preceding proposition, we know that $[a, b]$ is negligible. Thus, given any $\epsilon > 0$, there exists a sequence of open intervals (a_n, b_n) such that

$$[a, b] \subset \bigcup_{n=1}^{\infty} (a_n, b_n) \quad \text{and} \quad \sum_{n=1}^{\infty} (b_n - a_n) \leq \epsilon.$$

By the Heine-Borel theorem (4.5.4) there exists a positive integer r such that

$$[a, b] \subset (a_1, b_1) \cup \ldots \cup (a_r, b_r);$$

if we can infer from this inclusion that

$$(*) \qquad\qquad b - a < \sum_{k=1}^{r} (b_k - a_k),$$

it will follow that $b - a < \epsilon$, whence $b - a \leq 0$ by the arbitrariness of ϵ. The proof of $(*)$ is by induction on r. The case $r = 1$ is obvious. Let $r \geq 2$ and assume that the assertion has been proved for a covering of a closed interval by $r - 1$ open intervals. Rearranging the intervals (a_k, b_k) if necessary, we can suppose that $a \in (a_r, b_r)$, so that $a_r < a < b_r$.

If $b_r > b$ then $a_r < a \leq b < b_r$ and the inequality $(*)$ is obvious. On the other hand, if $b_r \leq b$ then

$$a_r < a < b_r \leq b,$$

therefore $[b_r, b]$ is disjoint from (a_r, b_r); however,

$$[b_r, b] \subset [a, b] \subset \bigcup_{k=1}^{r} (a_k, b_k),$$

and since $[b_r, b]$ is disjoint from the last term of the union, necessarily

$$[b_r, b] \subset \bigcup_{k=1}^{r-1} (a_k, b_k).$$

By the induction hypothesis,

$$b - b_r < \sum_{k=1}^{r-1} (b_k - a_k);$$

also $b_r - a < b_r - a_r$, and addition of the two inequalities yields $(*)$. $\Diamond$

For computational purposes, it will be useful to specify more sharply the coverings by intervals that figure in the definition of a negligible set (cf. the proof of 11.2.11); this involves some elementary observations on the 'algebra' of intervals:

11.1.7. **Proposition.** *Let* $\mathcal{A}$ *be the set of all finite unions*

$$A = \bigcup_{j=1}^{m} I_j \,,$$

where the I_j *are pairwise disjoint intervals in* $\mathbb{R}$ *(possibly degenerate, and not necessarily bounded). Then:*

(1) $A, B \in \mathcal{A} \Rightarrow A \cap B \in \mathcal{A}$,

(2) $A \in \mathcal{A} \Rightarrow \complement A \in \mathcal{A}$,

(3) $A, B \in \mathcal{A} \Rightarrow A \cup B \in \mathcal{A}$,

that is, $\mathcal{A}$ *is closed under finite intersections, complementation and finite unions. In particular,*

(4) $\mathcal{A}$ *coincides with the set of all finite unions of intervals (not necessarily pairwise disjoint).*

Proof. (1) Suppose $A = \bigcup_{j=1}^{m} I_j$ and $B = \bigcup_{k=1}^{n} J_k$, where the I_j (resp. the J_k) are pairwise disjoint intervals. Then

$$A \cap B = \bigcup_{j,k} I_j \cap J_k \,,$$

where the mn sets $I_j \cap J_k$ are intervals (4.1.5) and are pairwise disjoint, that is, if $j \neq j'$ or $k \neq k'$ then

$$(I_j \cap J_k) \cap (I_{j'} \cap J_{k'}) = (I_j \cap I_{j'}) \cap (J_k \cap J_{k'}) = \varnothing \,.$$

Thus $A \cap B \in \mathcal{A}$.

(2), (4) Note that if I is an interval then $\complement I$ is either a single interval (degenerate if $I = \mathbb{R}$) or the union of two disjoint intervals; for example,

$$\complement [a, +\infty) = (-\infty, a) \,, \quad \complement (a, b] = (-\infty, a] \cup (b, +\infty) \,.$$

In either case, $\complement I \in \mathcal{A}$. Thus, if $A = \bigcup I_j$ is any finite union of intervals (not necessarily pairwise disjoint) then $\complement A = \bigcap \complement I_j \in \mathcal{A}$ by (1). It follows that $A = \complement(\complement A) \in \mathcal{A}$. This proves both (2) and (4).

(3) This is immediate from (4), or from (1), (2) and the formula $A \cup B = \complement(\complement A \cap \complement B)$. ◇

11.1.8. **Corollary.** *If* $\mathcal{B}$ *is the set of all finite, disjoint unions of* **bounded** *intervals, then* $\mathcal{B}$ *is closed under the operations* $A \cap B$, $A \cup B$ *and* $A - B$.

Proof. If $A, B \in \mathcal{B}$ then the sets $A \cap B$, $A \cup B$ and $A - B = A \cap \complement B$ belong to the class $\mathcal{A}$ of the preceding proposition and are obviously bounded. $\Diamond$

11.1.9. Corollary. *If (I_n) is a sequence of closed intervals, then there exists a sequence of closed intervals (J_n) such that*
   (i) $\bigcup_{n=1}^{\infty} J_n = \bigcup_{n=1}^{\infty} I_n$,
   (ii) *the J_n are pairwise nonoverlapping,[3] and*
   (iii) *each J_n is a subinterval of some I_m.*

Proof. Let $A = \bigcup_{n=1}^{\infty} I_n$ and let A_n be the sequence of sets defined by

$$A_1 = I_1 \quad \text{and} \quad A_n = I_n - \bigcup_{k<n} I_k \quad \text{for } n > 1.$$

Writing $I_0 = \varnothing$, we have $A_n = I_n - \bigcup_{k<n} I_k$ for all $n \geq 1$.

Since $A_n \subset I_n$ for all n, we have $\bigcup A_n \subset \bigcup I_n = A$. In fact, $\bigcup A_n = A$; for, if $x \in A$ and n is the first index such that $x \in I_n$, then $x \in A_n$.

The sets A_n are pairwise disjoint; for, if $m < n$ and $x \in A_m$, then $x \in I_m$ and therefore $x \notin A_n$ by the definition of A_n.

Since every I_n belongs to the class $\mathcal{B}$ of the preceding corollary, it follows from that corollary that each A_n is a finite disjoint union of bounded intervals, necessarily subintervals of I_n. Say

$$A_n = \bigcup_{k=1}^{r_n} I_{nk},$$

where $I_{n1}, I_{n2}, \ldots, I_{nr_n}$ are pairwise disjoint subintervals of I_n; the closed interval J_{nk} having the same endpoints as I_{nk} is also a subinterval of I_n, and the intervals $J_{n1}, \ldots, J_{nr_n}$ are pairwise nonoverlapping.

To obtain the desired sequence $J_1, J_2, J_3, \ldots$, list the intervals $J_{11}, \ldots, J_{1n_1}$, then the intervals $J_{21}, \ldots, J_{2n_2}$, then the intervals $J_{31}, \ldots, J_{3n_3}$, and so on. Since the A_n are pairwise disjoint, the J_n are pairwise nonoverlapping. $\Diamond$

11.1.10. Corollary. *If N is a negligible set and $\epsilon > 0$, then there exists a sequence (J_n) of pairwise nonoverlapping closed intervals such that*

$$N \subset \bigcup J_n \quad \text{and} \quad \sum \ell(J_n) \leq \epsilon.$$

If, moreover, $N \subset [a,b]$, one can suppose that $J_n \subset [a,b]$ for all n.

Proof. Since N is negligible, there exists a sequence (I_n) of closed intervals such that $N \subset \bigcup I_n$ and $\sum \ell(I_n) \leq \epsilon$. Let (J_n) be the sequence

---

[3]Two intervals I and J are said to be *nonoverlapping* if their intersection is at most a single point. If the closures of I and J intersect in a single point, then I and J are said to *abut* each other.

of closed intervals constructed in the preceding corollary. With notations as in the proof of that corollary, $J_{n1}, \ldots, J_{nr_n}$ are pairwise nonoverlapping closed subintervals of I_n and it will suffice to show that

$$\ell(J_{n1}) + \ldots + \ell(J_{nr_n}) \leq \ell(I_n).$$

Changing notations, we are in the following situation: given nonoverlapping intervals $[a_1, b_1], \ldots, [a_r, b_r]$ such that

$$[a_1, b_1] \cup \ldots \cup [a_r, b_r] \subset [c, d],$$

we need only show that

(*)
$$\sum_{i=1}^{r} (b_i - a_i) \leq d - c.$$

Reordering the intervals $[a_i, b_i]$ if necessary (and eliminating any degenerate ones), we can suppose that $[a_i, b_i]$ is to the left of $[a_{i+1}, b_{i+1}]$ for $i = 1, \ldots, r - 1$, so that

$$c \leq a_1 \leq b_1 \leq a_2 \leq b_2 \leq \ldots \leq a_r \leq b_r \leq d;$$

the inequality (*) is then verified by a 'telescoping sum' argument:

$$d - c = (a_1 - c) + \sum_{i=1}^{r}(b_i - a_i) + \sum_{i=1}^{r-1}(a_{i+1} - b_i) + (d - b_r)$$

$$\geq 0 + \sum_{i=1}^{r}(b_i - a_i) + 0 + 0.$$

Finally, if $N \subset [a, b]$ then the intervals $J_n \cap [a, b]$ meet the requirements of the corollary. $\Diamond$

11.1.11. Definition. A statement is said to be true **almost everywhere** (briefly, 'a.e.') on a set $S \subset \mathbb{R}$ if there exists a negligible set N such that the statement is true for every x in $S - N$; this is also expressed by saying that the statement is true for **almost every** point of S (or for 'almost all' points of S).

11.1.12. Examples. (i) The statement 'F is differentiable a.e. on (a, b)' means that there exists a negligible set N such that $F'(x)$ exists for all $x \in (a, b) - N$.

(ii) The statement '$f \leq g$ a.e. on $[a, b]$' means that there exists a negligible set N such that $f(x) \leq g(x)$ for all $x \in [a, b] - N$.

(iii) See also Exercise 5.

Exercises

1. Let $f : [a, b] \to \mathbb{R}$ be the (discontinuous) function such that $f(x) = 1$ for $x \in (a, b)$ and $f(a) = f(b) = 0$. Show that f is «integrable» in the sense of the introduction.

2. Let $F : [0, 1] \to \mathbb{R}$ be the function defined by

$$F(x) = \begin{cases} x \sin(1/x) & \text{for } 0 < x \le 1 \\ 0 & \text{for } x = 0 \end{cases}$$

and define $f : [0, 1] \to \mathbb{R}$ by the formulas

$$f(x) = \begin{cases} F'(x) & \text{for } 0 < x \le 1 \\ 0 & \text{for } x = 0 . \end{cases}$$

Show that f is «integrable» in the sense of the introduction, but is not Riemann-integrable.

3. (Cantor set) For any closed interval $[a, b]$, $a < b$, take out the open middle third and write $r(\text{I})$ for what remains:

$$r(\text{I}) = [a, a + \frac{1}{3}(b - a)] \cup [b - \frac{1}{3}(b - a), b] .$$

More generally, if $A = \text{I}_1 \cup \ldots \cup \text{I}_n$ is a finite disjoint union of nondegenerate closed intervals, define

$$r(A) = r(\text{I}_1) \cup \ldots \cup r(\text{I}_n)$$

(which is the disjoint union of $2n$ closed intervals, the sum of whose lengths is $2/3$ of the sum of the lengths of the I_k). Define powers of r recursively by iterating this operation:

$$r^1(A) = r(A), \; r^2(A) = r(r(A)), \; r^{n+1}(A) = r(r^n(A)) .$$

The set

$$\Gamma = \bigcap_{n=1}^{\infty} r^n(\text{I}) ,$$

where $\text{I} = [0, 1]$, is called the *Cantor set*.[4] Prove:
(i) Γ is negligible.
(ii) Γ is uncountable, that is, no sequence (x_n) in Γ can exhaust Γ.
{Hints: (i) For each n, consider the covering of Γ by the intervals of $r^n(\text{I})$ and calculate their total length. (ii) Let I_1 be one of the intervals

---

[4] After Georg Cantor (1845–1918).

of $r(I)$ that excludes x_1, let I_2 be one of the intervals of $r(I_1)$ that excludes x_2, and so on (cf. Exercises 1 and 4 of §2.6).}

4. Write out the details for the proof of assertion (3) of Proposition 11.1.5.

5. Justify the statement that *almost every* real number is irrational.

11.2. Absolutely Continuous Functions

If $f : [a,b] \to \mathbb{R}$ is Riemann-integrable and $k = \sup |f|$, then the function $F : [a,b] \to \mathbb{R}$ defined by $F(x) = \int_a^x f$ satisfies the inequality $|F(x) - F(y)| \le k|x - y|$ for all x, y in $[a,b]$ (cf. 9.3.2). Functions F satisfying such an inequality have a name:

11.2.1. *Definition.* A function $F : [a,b] \to \mathbb{R}$ is said to satisfy a **Lipschitz condition**[1] (or to be a 'Lipschitz function') if there exists a constant M such that

$$|F(x) - F(y)| \le M|x - y|$$

for all x, y in $[a,b]$.

Not every Lipschitz function F is of the form $F(x) = F(a) + \int_a^x f$ for some Riemann-integrable function f; the proof is not easy.[2]

11.2.2. *Remark.* Every Lipschitz function is continuous. For, with notations as in the definition, it is clear that $|x_n - x| \to 0 \Rightarrow |F(x_n) - F(x)| \to 0$.

11.2.3. *Remark.* With notations as in the definition, if F is differentiable at a point $x \in (a,b)$ then $|F'(x)| \le M$ (because the absolute value of every 'difference quotient' of F is $\le M$). It is also true (but not easy to prove) that a Lipschitz function on $[a,b]$ is differentiable at almost every point of (a,b).[3]

11.2.4. Proposition. *The following conditions on a function $F : [a,b] \to \mathbb{R}$ are equivalent:*
(a) *F is Lipschitz;*

---

[1]Named after Rudolf Lipschitz (1832–1903).
[2]There exists a 'Lebesgue-integrable' (cf. 11.5.1 below) bounded function f on $[a,b]$ such that f is not equal almost everywhere to a Riemann-integrable function (see Example 32 on p.106 of B.R. Gelbaum and J.M.H. Olmsted's *Counterexamples in analysis* [Holden–Day, San Francisco, 1964]; the 'Lebesgue indefinite integral' F of f is a Lipschitz function not of the form $F(x) = F(a) + \int_a^x g$ for any Riemann-integrable function g.
[3]Cf. E. Hewitt and K. Stromberg, *Real and abstract analysis* [Springer–Verlag, New York, 1965], p. 267, (17.17).

(b) F *is continuous and there exists a constant* M *such that* $\ell\big(F(\mathrm{I})\big) \le$ $M\ell(\mathrm{I})$ *for every closed subinterval* I *of* $[a,b]$.

Proof. (a) $\Rightarrow$ (b): Let M be a constant such that $|F(x) - F(y)| \le$ $M|x - y|$ for all x, y in $[a, b]$ and let I be a closed subinterval of $[a, b]$. Since F is continuous (11.2.2), $F(\mathrm{I})$ is a closed interval (6.3.1), say $F(\mathrm{I}) = [F(c), F(d)]$, where $c, d \in \mathrm{I}$. Then

$$\ell\big(F(\mathrm{I})\big) = |F(c) - F(d)| \le M|c - d| \le M\ell(\mathrm{I}).$$

(b) $\Rightarrow$ (a): Given $x, y \in [a, b]$, let I be the closed subinterval of $[a, b]$ with endpoints x and y. Then $F(x), F(y) \in F(\mathrm{I})$, therefore

$$|F(x) - F(y)| \le \ell\big(F(\mathrm{I})\big) \le M\ell(\mathrm{I}) = M|x - y|,$$

thus F is Lipschitz. $\Diamond$

Every Lipschitz function maps negligible sets to negligible sets:

11.2.5. Proposition. *If* $F : [a, b] \to \mathbb{R}$ *is Lipschitz and* N *is a negligible subset of* $[a, b]$, *then* $F(\mathrm{N})$ *is a negligible subset of* $\mathbb{R}$.

Proof. Let $M > 0$ be a constant such that $|F(x) - F(y)| \le M|x - y|$ for all x, y in $[a, b]$. Given any $\epsilon > 0$, let (I_n) be a sequence of closed intervals such that $\mathrm{N} \subset \bigcup \mathrm{I}_n$ and $\sum \ell(\mathrm{I}_n) \le \epsilon/M$; replacing I_n by $\mathrm{I}_n \cap [a, b]$, we can suppose that $\mathrm{I}_n \subset [a, b]$. Then $F(\mathrm{N}) \subset \bigcup F(\mathrm{I}_n)$ and, by the preceding proposition, $\ell\big(F(\mathrm{I}_n)\big) \le M\ell(\mathrm{I}_n)$ for all n, therefore

$$\sum \ell\big(F(\mathrm{I}_n)\big) \le M \sum \ell(\mathrm{I}_n) \le M(\epsilon/M) = \epsilon;$$

thus, $F(\mathrm{N})$ can be covered by a sequence of intervals whose total length is as small as we like. $\Diamond$

The mapping property in this proposition is so central to our discussion that it is useful to have an abbreviation for it:

11.2.6. *Definition*. We shall say that a function $F : [a, b] \to \mathbb{R}$ is **negligent** if the image of every negligible subset of $[a, b]$ is negligible, that is,

$$\mathrm{N} \subset [a, b] \text{ negligible} \Rightarrow F(\mathrm{N}) \text{ negligible}.$$

Every Lipschitz function is negligent (11.2.5) but the converse is false; indeed, since a negligent function remains negligent if it is redefined at a finite number of points, a negligent function need not be continuous. An example of a continuous function that is not negligent is sketched in Exercise 2.

The class of Lipschitz functions is big enough to contain all of the indefinite integrals $F(x) = F(a) + \int_a^x f$ of Riemann-integrable functions f

(though it is not exhausted by them). On the other hand, the class is too small to contain all of the indefinite integrals associated with the 'Lebesgue-integrable' functions f (yet to be defined, in §11.5); what is required is a condition weaker than Lipschitz:

11.2.7. Definition. A function $F : [a, b] \to \mathbb{R}$ is said to be **absolutely continuous** (briefly, AC) if, given any $\epsilon > 0$, there exists a $\delta > 0$ such that, for finite lists $[a_1, b_1], \ldots, [a_n, b_n]$ of pairwise nonoverlapping closed subintervals of $[a, b]$,

$$\sum_{k=1}^{n}(b_k - a_k) \leq \delta \ \Rightarrow \ \sum_{k=1}^{n}|F(b_k) - F(a_k)| \leq \epsilon.$$

11.2.8. Remarks. Every Lipschitz function is absolutely continuous (with the notations of 11.2.1, given $\epsilon > 0$ let $\delta = \epsilon/M$), and every absolutely continuous function is continuous (in the notations of 11.2.7, consider $n = 1$). Thus,

$$\text{Lipschitz} \ \Rightarrow \ \text{absolutely continuous} \ \Rightarrow \ \text{continuous}.$$

Neither implication is reversible: there exists an absolutely continuous function that is not Lipschitz (Exercise 5), and a continuous function that is not absolutely continuous (cf. Exercise 2 and Proposition 11.2.11 below).

There is an analogue of Proposition 11.2.4 for absolute continuity:

11.2.9. Proposition. *The following conditions on a function $F : [a, b] \to \mathbb{R}$ are equivalent*:

(a) *F is absolutely continuous*;

(b) *F is continuous and, given any $\epsilon > 0$, there exists a $\delta > 0$ such that, for finite lists $I_1, \ldots, I_n$ of pairwise nonoverlapping closed subintervals of $[a, b]$*,

$$\sum_{k=1}^{n}\ell(I_k) < \delta \ \Rightarrow \ \sum_{k=1}^{n}\ell\big(F(I_k)\big) < \epsilon.$$

Proof. (a) $\Rightarrow$ (b): Given $\epsilon > 0$, choose $\delta > 0$ as in Definition 11.2.7. Let $I_1, \ldots, I_n$ be pairwise nonoverlapping closed subintervals of $[a, b]$ such that $\sum_{k=1}^{n}\ell(I_k) < \delta$. Say $F(I_k) = [F(c_k), F(d_k)]$, where $c_k, d_k \in I_k$. If J_k is the closed subinterval of I_k with endpoints c_k, d_k (in some order) then

$$\sum_{k=1}^{n}\ell(J_k) \leq \sum_{k=1}^{n}\ell(I_k) < \delta,$$

therefore $\sum_{k=1}^{n}|F(c_k) - F(d_k)| < \epsilon$ by the choice of δ, in other words $\sum_{k=1}^{n}\ell\big(F(I_k)\big) < \epsilon$.

(b) $\Rightarrow$ (a): For every closed subinterval I of $[a, b]$, $F(I)$ is a closed interval in $\mathbb{R}$ by the continuity of F (6.3.1). Given $\epsilon > 0$, choose $\delta > 0$

as in condition (b). Let $I_1, \ldots, I_n$ be nonoverlapping closed subintervals of $[a, b]$ such that $\sum \ell(I_k) < \delta$. Say $I_k = [a_k, b_k]$. Then $F(a_k), F(b_k) \in F(I_k)$, therefore $|F(a_k) - F(b_k)| \leq \ell(F(I_k))$, and

$$\sum_{k=1}^{n} |F(a_k) - F(b_k)| \leq \sum_{k=1}^{n} \ell(F(I_k)) < \epsilon$$

by the choice of δ. $\Diamond$

Linear combinations of absolutely continuous functions are also absolutely continuous (for products, see Exercise 6):

11.2.10. Proposition. *If F and G are absolutely continuous and c is a real number, then the functions $F + G$ and cF are also absolutely continuous.*

Proof. This follows easily from the relations

$$|(F + G)(b_i) - (F + G)(a_i)| \leq |F(b_i) - F(a_i)| + |G(b_i) - G(a_i)|,$$
$$|(cF)(b_i) - (cF)(a_i)| = |c|\,|F(b_i) - F(a_i)|,$$

where the a_i, b_i are endpoints of nonoverlapping intervals of total length $< \delta$, δ having been chosen in Definition 11.2.7 to 'work' for both F and G (take δ to be the minimum of a δ_1 that works for F and a δ_2 that works for G). $\Diamond$

The next proposition is the key to the 'uniqueness theorem' of §11.3 that makes possible the shortcut to the Lebesgue theory given in §11.5 (it also figures in the proof of Lebesgue's characterization of Riemann-integrability given in §11.4):

11.2.11. Proposition. *Every absolutely continuous function $F : [a, b] \to \mathbb{R}$ is negligent, that is,*

$$N \subset [a, b] \text{ negligible } \Rightarrow F(N) \text{ negligible.}$$

Proof. Let N be a negligible subset of $[a, b]$. Given any $\epsilon > 0$, choose $\delta > 0$ as in Definition 11.2.7. Let (I_n) be a sequence of closed intervals such that $N \subset \bigcup I_n$ and $\sum \ell(I_n) \leq \delta$. We can suppose that the I_n are nonoverlapping subintervals of $[a, b]$ (11.1.10). Since $F(N) \subset \bigcup F(I_n)$ and the $F(I_n)$ are closed intervals, it will suffice to show that $\sum \ell(F(I_n)) \leq \epsilon$.

Say $F(I_n) = [F(c_n), F(d_n)]$, where $c_n, d_n \in I_n$. If J_n is the closed subinterval of I_n with endpoints c_n and d_n (in some order) then

$$\sum \ell(J_n) \leq \sum \ell(I_n) \leq \delta,$$

therefore $\sum |F(c_n) - F(d_n)| \leq \epsilon$ by the choice of δ (the latter inequality is true for the partial sums of the series, hence for their supremum), in other words $\sum \ell(F(I_n)) \leq \epsilon$. $\Diamond$

It is possible to give an equivalent definition of absolute continuity in terms of the concept of negligence; the facts are reported in Exercise 4, but the proofs are not elementary.

Exercises

1. If f is any bounded real-valued function on a closed interval $[a, b]$, then each of the functions $F(x) = \overline{\int_a^x} f$ and $H(x) = \underline{\int_a^x} f$ is Lipschitz.

2. A continuous function need not be negligent.

{Sketch of proof: Let Γ be the Cantor set (§11.1, Exercise 3). There exists a continuous function $F_0 : \Gamma \to [0, 1]$ that is surjective,[4] and F_0 can be extended to a continuous function $F : [0, 1] \to [0, 1]$ by the Tietze extension theorem.[5] The Cantor set is negligible but $F(\Gamma) = [0, 1]$ is not.}

3. If F is negligent and $G = F$ a.e., it does not follow that G is negligent.

{Hint: Let $F : [0, 1] \to \mathbb{R}$ be the constant function $F(x) = 0$ and let $G : [0, 1] \to \mathbb{R}$ be a function such that $G(\Gamma) = [0, 1]$ and $G(x) = 0$ for $x \in [0, 1] - \Gamma$ (cf. the hint for Exercise 2).}

4. The following conditions on a function $F : [a, b] \to \mathbb{R}$ are equivalent:

(a) F is absolutely continuous;

(b) $F = G - H$, where G and H are continuous, strictly increasing and negligent;

(c) F is continuous, negligent and of bounded variation.

{A function $F : [a, b] \to \mathbb{R}$ is said to be of *bounded variation* if there is a finite upper bound for the sums $\sum_{k=1}^{n} |F(a_k) - F(b_k)|$, where $[a_1, b_1], \ldots,$ $[a_n, b_n]$ is any finite list of nonoverlapping closed subintervals of $[a, b]$.[6] Though the equivalences are easy enough to state, their proofs involve concepts and results not taken up in this book, and the proofs of some of the implications are downright difficult.[7]}

---

[4]Cf. J.L. Kelley, *General topology* [Van Nostrand, Princeton, 1955], p. 166, (e).

[5]Cf. J. Dixmier, *General topology* [Springer–Verlag, New York, 1984], p. 85, 7.6.1.

[6]An equivalent condition is that $F = G - H$ with G and H increasing functions [cf. E. Hewitt and K. Stromberg, *op. cit.*, p. 266, (17.16)].

[7]The equivalence (a) $\Leftrightarrow$ (c) is due to S. Banach [cf. Hewitt and Stromberg, *op. cit.*, p. 288, (18.25)]. For a proof of (a) $\Leftrightarrow$ (b), see the author's article in *Paul Halmos: Celebrating 50 years of mathematics* [Springer–Verlag, New York, 1991], p. 284, Proposition.

5. The function $F : [0,1] \to \mathbb{R}$ defined by $F(x) = \sqrt{x}$ is absolutely continuous but not Lipschitz.

{Sketch of proof:[8] Since F has a derivative on $(0,1)$ that is unbounded, F is not a Lipschitz function. That F is absolutely continuous can be inferred from the fact that $F(x)$ is the 'Lebesgue integral' from 0 to x of the function $f : [0,1] \to \mathbb{R}$ defined by $f(x) = \frac{1}{2}x^{-1/2}$ for $0 < x \le 1$ and $f(0) = 0$.}

6. If F and G are absolutely continuous functions on $[a,b]$, then so is FG.

{Hint: Let M and N be the least upper bounds (i.e., the maximum values) of the functions $|F|$ and $|G|$, respectively, and consider the identity

$$F(c)G(c) - F(d)G(d) = F(c)[G(c) - G(d)] + [F(c) - F(d)]G(d),$$

whose right side has absolute value $\le M|G(c) - G(d)| + N|F(c) - F(d)|$.}

7. In the context of Riemann-Stieltjes integration (§9.1, Exercise 5) let F and H be the indefinite upper and lower RS-integrals of a bounded function f with respect to an increasing function g (§9.3, Exercise 5). If g is absolutely continuous (or Lipschitz) then the same is true of F and H.

11.3. The Uniqueness Theorem

The core result of this section is a refinement of an argument in the proof of Theorem 8.5.5:

11.3.1. Lemma. *Let* $F : [a,b] \to \mathbb{R}$ *be a continuous function and let* N *be a subset of* $\mathbb{R}$ *such that, for every* $x \in (a,b) - N$, F *is right differentiable at* x *and* $F'_r(x) > 0$.

(i) *If* $F(N)$ *has empty interior then* F *is an increasing function.*

(ii) *If both* N *and* $F(N)$ *have empty interior, then* F *is strictly increasing.*

Proof.[1] (i) Let $a \le c < d \le b$. We are to show that $F(c) \le F(d)$.

Assume to the contrary that $F(d) < F(c)$. Since $F(N)$ has empty interior, it cannot contain the nondegenerate open interval $(F(d), F(c))$; choose a point $k \in (F(d), F(c))$ such that $k \notin F(N)$. Thus,

$$F(d) < k < F(c) \text{ and, for all } x \in N, \ F(x) \ne k.$$

[8]Cf. K. Stromberg, *An introduction to classical real analysis* [Wadsworth, Belmont, CA, 1981], p. 163.

[1]Inspired by an argument in E.J. McShane's *Integration* [Princeton University Press, 1944], p. 200, 34.1.

By the intermediate value theorem (6.1.2), F assumes the value k at some point of the interval $[c,d]$, that is, the set

$$A = \{x \in [c,d] : F(x) = k\}$$

is nonempty. If $x_n \in A$ and $x_n \to x$, then $x \in [c,d]$ (because $[c,d]$ is a closed set) and $F(x_n) \to F(x)$ (because F is continuous); since $F(x_n) = k$ for all n, it follows that $F(x) = k$, thus $x \in A$. This shows that A is a closed set.

Let $s = \sup A$; then $s \in A$ (4.5.7), thus $F(s) = k$. So to speak, s is the right-most point of $[c,d]$ where F takes on the value k; since

$$F(d) < k = F(s) < F(c),$$

necessarily $d \neq s$ and $s \neq c$, thus $c < s < d$. Since F does not take on the value k in N, necessarily $s \notin$ N, therefore $s \in (a,b) - $N; by the hypothesis on N, F is right differentiable at s and $F_r'(s) > 0$.

Let (t_n) be a sequence such that $s < t_n < d$ and $t_n \to s$. From $t_n > s = \sup A$ we have $t_n \notin A$, thus $F(t_n) \neq k$. Necessarily $F(t_n) < k$ for all n; for, if $F(t_n) > k$ for some n, then the inequalities $F(d) < k < F(t_n)$ would imply the existence of a point in the open interval (t_n, d) at which F takes on the value k, contradicting the maximality of s. Thus $F(t_n) < k = F(s)$ for all n, and passage to the limit in the inequalities

$$\frac{F(t_n) - F(s)}{t_n - s} < 0$$

yields $F_r'(s) \leq 0$, contrary to the fact that $s \in (a,b) - $N.

(ii) Now assume that both N and $F($N$)$ have empty interior. By (i), F is increasing. Given $a \leq c < d \leq b$, we are to show that $F(c) < F(d)$. At any rate, $F(c) \leq F(d)$. Assume to the contrary that $F(c) = F(d)$. Then F is constant on $[c,d]$. In particular, if $x \in (c,d)$ then $F'(x) = 0$, therefore x cannot belong to $(a,b) - $N; it follows that $x \in$ N, and we have shown that $(c,d) \subset$ N, contrary to the assumption that N has empty interior. $\Diamond$

11.3.2. Theorem. *If $F : [a,b] \to \mathbb{R}$ is absolutely continuous and if, at almost every point of (a,b), F has a right derivative > 0, then F is strictly increasing.*

Proof.[2] By assumption, there exists a negligible set $N \subset [a,b]$ such that, at every point $x \in (a,b) - $N, F is right differentiable and $F_r'(x) > 0$.

---

[2]It can be shown that an absolutely continuous function is differentiable almost everywhere [cf. Hewitt and Stromberg, *Real and abstract analysis* [Springer-Verlag, New York, 1965], p. 267, (17.17) and p. 283, (18.12); in the present theorem we are not begging the question of differentiability, but are *assuming* that F is right differentiable almost everywhere.

Since F is negligent (11.2.11), $F(N)$ is also negligible. Thus both N and $F(N)$ have empty interior (11.1.6) and the hypotheses of (ii) of the lemma are fulfilled. ◇

11.3.3. Corollary. *If $F : [a,b] \to \mathbb{R}$ is absolutely continuous and if, at almost every point of (a,b), F has a right derivative ≥ 0, then F is increasing.*

Proof. Assuming $a \leq c < d \leq b$, we are to show that $F(c) \leq F(d)$. Given any $\epsilon > 0$, it will suffice to show that $F(c) < F(d) + \epsilon(d-c)$.

The function $G : [a,b] \to \mathbb{R}$ defined by $G(x) = F(x) + \epsilon x$ is absolutely continuous (11.2.10) and, at almost every point of (a,b), G is right differentiable with

$$G'_r(x) = F'_r(x) + \epsilon \geq \epsilon > 0,$$

therefore G is strictly increasing by the preceding corollary; in particular, $G(c) < G(d)$, that is,

$$F(c) + \epsilon c < F(d) + \epsilon d,$$

whence the desired inequality. ◇

11.3.4. Theorem. (**Uniqueness theorem**) *If $F : [a,b] \to \mathbb{R}$ is absolutely continuous and if, at almost every point of (a,b), F has a right derivative equal to 0, then F is a constant function.*

Proof. Both F and $-F$ satisfy the hypotheses of the preceding corollary, consequently F and $-F$ are both increasing; in other words, F is both increasing and decreasing, hence constant. ◇

The following application makes a nice connection with Riemann-integrability (to be exploited more fully in the next section):

11.3.5. Corollary. *If $f : [a,b] \to \mathbb{R}$ is a bounded function that has a right limit at almost every point of (a,b), then f is Riemann-integrable.*

Proof. Let F and H be the indefinite upper and lower integrals of f (9.3.1), that is,

$$F(x) = \overline{\int_a^x} f \quad \text{and} \quad H(x) = \underline{\int_a^x} f$$

for all $x \in [a,b]$. Then F and H are Lipschitz functions (9.3.2) and, for almost every x in (a,b),

$$F'_r(x) = f(x+) = H'_r(x)$$

by the proof of Theorem 9.6.11. Thus $F - H$ is absolutely continuous (11.2.8) and has right derivative 0 at almost every point of (a,b), consequently $F - H$ is constant by the uniqueness theorem (11.3.4). Inasmuch as $F(a) = H(a) = 0$, we conclude that $F - H$ is the function identically zero; in particular $F(b) = H(b)$, that is, $\overline{\int_a^b} f = \underline{\int_a^b} f$. ◊

Exercises

1. What about left derivatives?
{Hint: Adapt the argument of 11.3.1 by considering instead $s = \inf A$.}

2. Every theorem in Chapter 9 affirming the Riemann-integrability of a class of functions is a special case of Corollary 11.3.5 (cf. 9.4.4, 9.4.5, 9.6.5, 9.6.7, 9.6.11 and 9.6.12).

11.4. Lebesgue's Criterion for Riemann-Integrability

Here it is:

11.4.1. Theorem. *The following conditions on a bounded function* $f : [a, b] \to \mathbb{R}$ *are equivalent:*
(a) f *is Riemann-integrable;*
(b) f *is continuous at almost every point of* $[a, b]$, *that is, the set*

$$D = \{x \in [a, b] : \ f \ \text{is not continuous at } x\}$$

of points of discontinuity of f *is negligible.*

The implication (b) ⇒ (a) follows at once from Corollary 11.3.5; indeed, one need only assume that f has a right limit at almost every point of (a, b) (not a true generalization, since the reverse implication will ensure continuity almost everywhere). The proof of the reverse implication requires some preparation.

11.4.2. Lemma. *If* $f : [a, b] \to \mathbb{R}$ *is a bounded function, c is a point of* $[a, b]$, *and*

$$S_c = \{y \in \mathbb{R} : \ f(x_n) \to y \ \text{for some sequence } x_n \to c\},$$

then S_c *is a closed and bounded subset of* $\mathbb{R}$ *with* $f(c) \in S_c$.

Proof. Suppose $y_n \in S_c$ and $y_n \to y$; we are to show that $y \in S_c$. For each n, it follows from the definition of S_c that there exists a point $x_n \in [a, b]$ such that $|x_n - c| < 1/n$ and $|f(x_n) - y_n| < 1/n$. Then $x_n \to c$ and, since

$$|f(x_n) - y| \le |f(x_n) - y_n| + |y_n - y| \to 0 + 0,$$

we have $f(x_n) \to y$, thus $y \in S_c$.

Since f is a bounded function, the set S_c is bounded; indeed, if $k \leq f(x) \leq K$ for all $x \in [a, b]$, then $S_c \subset [k, K]$. Finally, $f(c) \in S_c$ (consider the sequence $x_n = c$ for all n). ◊

11.4.3. Definition. Let $f : [a, b] \to \mathbb{R}$ be a bounded function and, for each point $c \in [a, b]$, let S_c be the set defined in the lemma. Since S_c is nonempty, closed and bounded, it has a smallest and a largest element (4.5.7). We define two functions $\underline{f}, \overline{f} : [a, b] \to \mathbb{R}$ by the formulas

$$\underline{f}(c) = \min S_c, \quad \overline{f}(c) = \max S_c \quad \text{for all } c \in [a, b].$$

Clearly $\underline{f}$ and $\overline{f}$ are bounded (by the same bounds as f) and $\underline{f} \leq f \leq \overline{f}$ on $[a, b]$ (because $f(c) \in S_c$).

11.4.4. Lemma. *With notations as in the definition, let $c \in [a, b]$. Then:*

$$f \text{ is continuous at } c \quad \Leftrightarrow \quad \underline{f}(c) = \overline{f}(c),$$

in which case $\underline{f}(c) = f(c) = \overline{f}(c)$.

Proof. $\Rightarrow$: If $x_n \to c$ then $f(x_n) \to f(c)$ by the continuity of f at c, therefore $S_c = \{f(c)\}$; thus the elements $\underline{f}(c)$ and $\overline{f}(c)$ of S_c coincide with its unique element $f(c)$.

$\Leftarrow$: If $\underline{f}(c) = \overline{f}(c)$ then $\underline{f}(c) = f(c) = \overline{f}(c)$ and $S_c = \{f(c)\}$. Given any sequence (x_n) in $[a, b]$ such that $x_n \to c$, we are to show that $f(x_n) \to f(c)$. Thus, given any $\epsilon > 0$, we must show that $|f(x_n) - f(c)| < \epsilon$ ultimately.

Assume to the contrary that $|f(x_n) - f(c)| \geq \epsilon$ frequently. Passing to a subsequence, we can suppose that $|f(x_n) - f(c)| \geq \epsilon$ for all n. Since the sequence $(f(x_n))$ is bounded, by the Weierstrass–Bolzano theorem it has a convergent subsequence; passing again to a subsequence, we can suppose that $f(x_n) \to y$ for some real number y. Then $y \in S_c = \{f(c)\}$, therefore $y = f(c)$; thus $f(x_n) \to f(c)$ and passage to the limit in the inequalities $|f(x_n) - f(c)| \geq \epsilon$ yields the absurdity $|f(c) - f(c)| \geq \epsilon$. ◊

11.4.5. Lemma. *If $f : [a, b] \to \mathbb{R}$ is a bounded function and $a < c < b$, then*

$$0 \leq \overline{f}(c) - \underline{f}(c) \leq M - m,$$

where (as in §9.1) $M = \sup f$ *and* $m = \inf f$.

Proof. If $x_n \to c$ and $f(x_n) \to y$ then $m \leq f(x_n) \leq M$ for all n, and passage to the limit yields $m \leq y \leq M$. Thus $S_c \subset [m, M]$ and, in particular,

$$m \leq \underline{f}(c) \leq \overline{f}(c) \leq M,$$

whence the asserted inequalities. ◊

We can now complete the proof of Theorem 11.4.1.

Proof of (a) $\Rightarrow$ (b): Assuming that $f : [a, b] \to \mathbb{R}$ is Riemann-integrable, we are to show that the set D of points of discontinuity of f is negligible. By Lemma 11.4.4,

$$D = \{x \in [a, b] : \overline{f}(x) - \underline{f}(x) > 0\}.$$

For each positive integer k, let

$$D_k = \{x \in [a, b] : \overline{f}(x) - \underline{f}(x) > 1/k\};$$

since $D = \bigcup_{k=1}^{\infty} D_k$, it will suffice to show that each D_k is negligible (11.1.5).

Fix an index k and let $\epsilon > 0$. Since f is Riemann-integrable, by Theorem 9.7.5 there exists a subdivision

$$\sigma = \{a = a_0 < a_1 < \cdots < a_n = b\}$$

of $[a, b]$ such that $W_f(\sigma) \leq \epsilon/k$, where k is the index just fixed and $W_f(\sigma)$ is the weighted oscillation of f for σ,

$$W_f(\sigma) = S(\sigma) - s(\sigma) = \sum_{\nu=1}^{n}(M_\nu - m_\nu)(a_\nu - a_{\nu-1})$$

(with notations as in §9.7). Since the negligibility of D_k is not influenced by the inclusion or omission of the finitely many points a_ν, it will suffice to show that the points of D_k not equal to any a_ν are all contained in the union of finitely many intervals of total length $< \epsilon$. Here are the intervals: let $I_1, \ldots, I_r$ be a faithful listing (no repetitions) of those intervals $(a_{\nu-1}, a_\nu)$ that contain at least one point of D_k; we need only show that $\sum_{j=1}^{r} \ell(I_j) < \epsilon$.

Fix an index $j \in \{1, \ldots, r\}$. Suppose $I_j = (a_{\nu-1}, a_\nu)$ and choose a point c of D_k that belongs to I_j. Since $c \in D_k$ we have

$$1/k < \overline{f}(c) - \underline{f}(c) \leq M_\nu - m_\nu,$$

where the latter inequality follows from applying the preceding lemma to the restriction of f to the interval $[a_{\nu-1}, a_\nu]$ (note that if x_n is a sequence in $[a, b]$ converging to c, then $x_n \in (a_{\nu-1}, a_\nu)$ ultimately). It follows that

(∗) $$\frac{1}{k}(a_\nu - a_{\nu-1}) < (M_\nu - m_\nu)(a_\nu - a_{\nu-1}).$$

Summing the inequalities (∗) over those ν for which $(a_{\nu-1}, a_\nu)$ contains at least one point of D_k, the sum of the left sides is

$$\frac{1}{k}\sum_{j=1}^{r} \ell(I_j),$$

whereas the sum of the right sides is $\leq W_f(\sigma)$, which was chosen to be $\leq \epsilon/k$, thus

$$\frac{1}{k}\sum_{j=1}^{r}\ell(\mathrm{I}_j) < \epsilon/k,$$

whence $\sum_{j=1}^{r}\ell(\mathrm{I}_j) < \epsilon$, as we wished to show. $\Diamond$

Lebesgue's criterion brings with it a fairly satisfying 'Fundamental theorem of calculus' for the Riemann integral:

11.4.6. Theorem. *Let* $f : [a, b] \to \mathbb{R}$ *be a Riemann-integrable function. Then:*

(1) *The function* $F : [a, b] \to \mathbb{R}$ *defined by*

$$F(x) = \int_a^x f \quad \text{for all } x \in [a, b]$$

is Lipschitz and $F'(x) = f(x)$ *for almost every* x *in* (a, b).

(2) *If* G *is any absolutely continuous function such that* $G' = f$ *a.e. then*

$$G(x) = G(a) + \int_a^x f \quad \text{for all } x \in [a, b];$$

in particular, G *is Lipschitz and*

$$\int_a^b f = G(b) - G(a).$$

Proof. (1) This is shown by the proof of 11.3.5.

(2) The function $G - F$ is absolutely continuous (11.2.10) and $(G - F)' = 0$ a.e., therefore $G - F$ is constant by the uniqueness theorem of the preceding section; thus, for all $x \in [a, b]$,

$$G(x) - F(x) = G(a) - F(a) = G(a) - 0,$$

and

$$G(x) = G(a) + F(x) = G(a) + \int_a^x f. \Diamond$$

The only blemish on this otherwise appealing result is that it offers no succinct description of the class of functions that can play the role of G (for all possible Riemann-integrable functions f); as noted following Definition 11.2.1, 'Lipschitz' is not enough. For complete satisfaction in this regard, we must advance to the Lebesgue integral.

Exercises

1. If f and g are Riemann-integrable then so is fg (§9.7, Exercise 4). Give an alternate proof based on Lebesgue's criterion.
{Hint: The union of two negligible sets is negligible.}

2. If a bounded function $f : [a, b] \to \mathbb{R}$ has a right limit at almost every point of $[a, b]$, then f is continuous at almost every point of $[a, b]$.

3. Let $f : [a, b] \to \mathbb{R}$ be Riemann-integrable, $g : [a, b] \to \mathbb{R}$ an increasing function, and write

$$F(x) = \overline{\int}_a^x f \, dg \quad \text{and} \quad H(x) = \underline{\int}_a^x f \, dg \quad (a \le x \le b)$$

for the Riemann-Stieltjes indefinite upper and lower integrals of f with respect to g (cf. §9.1, Exercise 5). Take it on faith that g is differentiable at almost every point of (a, b).[1] Then:

(i) At almost every x in (a, b), F and H are differentiable with $F'(x) = f(x)g'(x) = H'(x)$. {Hint: §9.3, Exercise 5.}

(ii) If the points of discontinuity of f can be listed in a sequence, and if g is continuous on $[a, b]$ and the points of (a, b) where g is not differentiable can be listed in a sequence, then f is RS-integrable with respect to g. {Hint: Apply §8.5, Exercise 10 to $F - H$.}

(iii) If there exists a point c in $[a, b)$ such that neither f nor g is right continuous at c, then f is not RS-integrable with respect to g.[2]

11.5. Lebesgue-Integrable Functions

Early in the twentieth century, the French mathematician Henri Lebesgue (1875–1941), beginning with his doctoral dissertation, introduced and perfected a theory of integration applicable to a class of functions nowadays called 'Lebesgue-integrable'.[3] A summit of the theory is his characterization of the class of 'indefinite integrals' associated with these functions: it is precisely the class of absolutely continuous functions.

Setting aside for the moment (as regards this book, forever) his precise definition of 'integrability', which entails a lot of technical machinery,[4]

---

[1]Cf. E. Hewitt and K. Stromberg, *Real and abstract analysis* [Springer-Verlag, New York, 1965], p. 264, (17.12).

[2]Cf. T. H. Hildebrandt, *Introduction to the theory of integration* [Academic Press, New York, 1963], p. 50, 10.6.

[3]Cf. H. Lebesgue, *Leçons sur l'intégration et la recherche des fonctions primitives* [Gauthier–Villars, 2nd edn., Paris, 1928], T. Hawkins, *Lebesgue's theory of integration: Its origins and development* [2nd edn., Chelsea, New York, 1979].

[4]Cf. B. Sz.-Nagy, *Introduction to real functions and orthogonal expansions* [Oxford University Press, New York, 1965], Chapter 5.

Lebesgue gave an astonishingly simple characterization of such functions:
A function $f : [a, b] \to \mathbb{R}$ is 'Lebesgue-integrable' if and only if there exists
an absolutely continuous function $F : [a, b] \to \mathbb{R}$ such that $F'(x) = f(x)$
for almost every x in (a, b) .[5]

There is more to the story, as we shall note at the end of the section;
for the present, we take Lebesgue's characterization as a cue for giving an
alternative, elementary *definition* of Lebesgue-integrability:

11.5.1. *Definition.* We shall say that a function $f : [a, b] \to \mathbb{R}$ is
Lebesgue-integrable if there exists an absolutely continuous function
$F : [a, b] \to \mathbb{R}$ such that $F'(x) = f(x)$ for almost every x in (a, b)
(briefly, $F' = f$ a.e.). Such a function F is called a **primitive** for f.
{Caution: It is not enough that $F' = f$ a.e.; we also require that a
primitive be absolutely continuous (cf. Exercise 5).}

If G is another primitive for f, then $F - G$ is a constant function
(11.3.4); in particular,

$$(F - G)(b) = (F - G)(a) ,$$

so that $F(b) - F(a) = G(b) - G(a)$. Thus, there is no ambiguity in the
following definition:

11.5.2. *Definition.* With notations as in the preceding definition, the
number $F(b) - F(a)$ (which is independent of the choice of a particular
primitive F) is called the **Lebesgue integral** of f from a to b, written

$$\int_a^b f = F(b) - F(a) .$$

11.5.3. *Example.* By Theorem 11.4.6, every Riemann-integrable function
is Lebesgue-integrable and its integral as given by the preceding definition
coincides with its Riemann integral; thus the use of the same symbol $\int_a^b f$
in the two contexts is consistent.

11.5.4. *Example.* The function $f : [a, b] \to \mathbb{R}$ defined by

$$f(x) = \begin{cases} 1 & \text{for } x \text{ rational} \\ 0 & \text{for } x \text{ irrational} \end{cases}$$

is not Riemann-integrable (by either 9.4.3 or 11.4.1), but f is Lebesgue-
integrable with Lebesgue integral 0.
{Proof: If F is any constant function, then $F'(x) = 0 = f(x)$ for
every x in the set of irrationals of (a, b), a set whose complement in

---

[5]The corresponding statement with "Lebesgue" replaced by "Riemann", and
"absolutely continuous" by "Lipschitz" is false, as noted following 11.2.1.

$[a, b]$ is negligible (cf. §2.6, Exercise 1 and 11.1.3); since F is absolutely continuous, f is Lebesgue-integrable and $\int_a^b f = F(b) - F(a) = 0$.}

11.5.5. Remark. Let $f : [a, b] \to \mathbb{R}$ be Lebesgue-integrable and let F be a primitive for f. If $a < x \le b$ then the restriction of f to $[a, x]$ is also Lebesgue-integrable (the restriction of F to $[a, x]$ serves as a primitive) and its Lebesgue integral is $F(x) - F(a)$; stretching the notation (as was done for the Riemann integral in Definition 9.2.2) we write $\int_a^x f$ for this integral, and, with the convention $\int_a^a f = 0$, we have

$$F(x) - F(a) = \int_a^x f \quad \text{for all } x \in [a, b].$$

Thus, every primitive of f differs from the 'indefinite integral' function $x \mapsto \int_a^x f$ by a constant.

11.5.6. Remark. It can be shown that an absolutely continuous function $F : [a, b] \to \mathbb{R}$ has a derivative at almost every x in (a, b).[6] If $f : [a, b] \to \mathbb{R}$ is then defined by

$$f(x) = \begin{cases} F'(x) & \text{if it exists} \\ 0 & \text{otherwise} \end{cases}$$

we have $F' = f$ a.e., consequently f is Lebesgue-integrable and F is a primitive for f. In view of the preceding remark, it follows that every absolutely continuous function F has the form

$$F(x) = F(a) + \int_a^x f \quad (a \le x \le b)$$

for a suitable Lebesgue-integrable function f.

The price we pay for the ease of the *definition* of Lebesgue integral given above is the difficulty in proving its *properties*. Linearity comes effortlessly (Exercise 1) and a few other accessible properties are noted in the exercises. On the other hand, it is a fact that if f is Lebesgue-integrable then so is $|f|$, but it is not clear how to prove it via Definition 11.5.1. The situation is equally bleak as regards the 'convergence theorems' of the Lebesgue theory. For example, if g and f_n ($n = 1, 2, 3, \ldots$) are Lebesgue-integrable functions on $[a, b]$ such that $|f_n| \le |g|$ a.e. for every n, and if $f : [a, b] \to \mathbb{R}$ is a function such that $f_n(x) \to f(x)$ for almost every x in $[a, b]$, then it is known that f is Lebesgue-integrable and that $\int_a^b f_n \to \int_a^b f$ (this

---

[6]The proof is not easy; cf. Hewitt and Stromberg [*op. cit.*, p. 283, (18.12)], B. Sz-Nagy [*op. cit.*, p. 94, Theorem, and pp. 108–111, Item 2], or the author's article in *Paul Halmos: Celebrating 50 years of mathematics* [Springer-Verlag, New York, 1991], p. 267, last paragraph of the Introduction.

result is known as the 'dominated convergence theorem'[7]); one shudders at
the prospect of inferring this result from Definition 11.5.1.

What we have given here is a painless overview of the 'answer' while
avoiding the hard technical considerations that are needed to get there;
the reader who is intrigued by the symmetry and elegance of the Lebesgue
theory will want to go on to overcome its technical challenges.

Exercises

1. If f and g are Lebesgue-integrable functions on $[a, b]$ and c is a
real number, then $f + g$ and cf are also Lebesgue-integrable and

$$\int_a^b (f + g) = \int_a^b f + \int_a^b g, \quad \int_a^b (cf) = c \int_a^b f.$$

2. If $f : [a, b] \to \mathbb{R}$ and $a < c < b$, then f is Lebesgue-integrable
on $[a, b]$ if and only if its restrictions to $[a, c]$ and $[c, b]$ are Lebesgue-
integrable, in which case

$$\int_a^b f = \int_a^c f + \int_c^b f$$

(see 11.5.5 for the notations).

3. Let $f : [a, b] \to \mathbb{R}$ be Lebesgue-integrable and let F be a primitive
for f, so that $F(x) = F(a) + \int_a^x f$ for all x in $[a, b]$ (11.5.5). Then:
  (i) $f \geq 0$ a.e. $\Leftrightarrow$ F is increasing.
  (ii) $f = 0$ a.e. $\Leftrightarrow$ F is constant.
  (iii) $f > 0$ a.e. $\Rightarrow$ F is strictly increasing.
  (iv) The converse of (iii) is false.[8]

4. If, for the definition of absolute continuity, we take instead the criterion (b) of §11.2, Exercise 4, then the entire theory of the Lebesgue integral
presented here rests effectively on one new concept beyond the preceding
chapters: the concept of negligible set. However, inspection of the proof
of the uniqueness theorem (especially 11.3.2 and 11.3.3) shows that it is
necessary to prove that (i) the functions of the form $F = G - H$, where
G and H are continuous, strictly increasing and negligent, are themselves
negligent, and that (ii) the set of all such functions F is closed under sums
and scalar multiples. The proofs of (i) and (ii) are elementary but tricky.[9]

---

[7]Cf. Hewitt and Stromberg [*op. cit.*, p. 172, (12.24)].

[8]Cf. H.L. Royden, *Real analysis* [3rd edn., Macmillan, New York, 1988],
p. 111, Exer. 19.

[9]Cf. the author's earlier-cited article in *Paul Halmos: Celebrating 50 years of
mathematics*, p. 283, Corollary 1, and his article "Why there is no 'Fundamental
theorem of calculus' for the Riemann integral" [*Expositiones Mathematicae* **11**
(1993), 271-279].

5. The constant function 0 on $[a, b]$ is Lebesgue-integrable and its primitives are the constant functions. However, there exists[10] a nonconstant, continuous increasing function $F : [a, b] \to \mathbb{R}$ such that $F'(x) = 0$ for almost every x; such a function cannot be absolutely continuous, or even negligent (cf. § 11.2, part (c) of Exercise 4).

6. Let $f : [a, b] \to \mathbb{R}$ be Riemann-integrable, $g : [a, b] \to \mathbb{R}$ an absolutely continuous increasing function, and h the function on $[a, b]$ defined by

$$h(x) = \begin{cases} f(x)g'(x) & \text{if } g \text{ is differentiable at } x, \\ 0 & \text{otherwise.} \end{cases}$$

Then:

(i) f is RS-integrable with respect to g.[11]

(ii) h is Lebesgue-integrable and $\int_a^b h = \int_a^b f \, dg$; stretching the notations a little, we can express this formula as $\int_a^b f \, dg = \int_a^b fg'$.[12]

{Hint: Contemplate §11.4, Exercise 3; §11.2, Exercise 7; and the uniqueness theorem of §11.3.}

---

[10]Cf. Hewitt and Stromberg [*op. cit.*, p. 113, (8.28)].

[11]Cf. E. W. Hobson, *The theory of functions of a real variable and the theory of Fourier series. Vol. 1*, p. 545 [3rd edn., Cambridge University Press, 1927; reprinted by Dover, New York, 1957].

[12]Cf. R. L. Jeffery, *The theory of functions of a real variable*, p. 206, Theorem 8.5 [2nd edn., University of Toronto Press, 1953].

Appendix

§A.1. Proofs, logical shorthand
§A.2. Set notations
§A.3. Functions
§A.4. Integers

A.1. Proofs, Logical Shorthand

A.1.1. A **proposition** is a statement which is either true or false, but not both. {Example of a proposition: "The integer 12,537,968 is divisible by 17." (Is it true?)} In the following discussion, letters $P, Q, R, \ldots$ represent propositions.

If P is a proposition, then the **negation** of P is the proposition, denoted $\sim P$ (or P'), that is true when P is false, and false when P is true. {Example: If P is the proposition "Every integer is ≥ 0" then $\sim P$ is the proposition "Some integer is < 0".}

A.1.2. Some useful shorthand for expressing propositions and relations between them:

SYMBOL	MEANING	EXAMPLES
&	and	$x > 0 \ \& \ x \leq 3$
		(also written $0 < x \leq 3$)
or	or	$x \geq 0$ or $x \leq 0$
$\Rightarrow$	implies	$P \Rightarrow Q$ (if P is true
		then Q must be true)
		n odd $\Rightarrow n+1$ even
$\Leftrightarrow$	if and only if	$P \Leftrightarrow Q$ means that both
		$P \Rightarrow Q$ and $Q \Rightarrow P$;
		$(x \geq 0 \ \& \ x \leq 0) \Leftrightarrow x = 0$
$\forall$	for all	$x^2 \geq 0$ ($\forall$ integers x)
$\exists$	there exists	
		$\exists \ x \ni x^2 = 4$
$\ni$	such that	
$\exists!$	there exists one	$\exists! \ x \ni 2x = 6$
	and only one	

For $P \Leftrightarrow Q$ we may also write $P \equiv Q$ (especially if P and Q themselves involve implication symbols).

220

A.1.3. A **theorem** typically consists of an implication $H \Rightarrow C$; one calls H the **hypothesis** of the theorem, C its **conclusion**. A **proof** of the theorem consists in verifying that **if** the hypothesis is true **then** the conclusion must also be true. The argument is sometimes broken down into a chain of several implications that are easier to verify, the conclusion of each being the hypothesis of the next (if there is a next); for example,

$$H \Rightarrow P_1 \Rightarrow P_2 \Rightarrow P_3 \Rightarrow C.$$

The mechanism for inferring $H \Rightarrow C$ from such a chain is the

Law of syllogism: If $P \Rightarrow Q$ and $Q \Rightarrow R$ then $P \Rightarrow R$.

Some other logical maneuvers permitted in proofs:

Law of the excluded middle: P or $\sim P$ (if P is not true then its negation $\sim P$ must be true).

Law of contradiction: $P \,\&\, (\sim P)$ is false (P and $\sim P$ can't be simultaneously true).

De Morgan's laws: $\sim (P\&Q) \equiv (\sim P)$ or $(\sim Q)$. That is, to say that P and Q are not both true is to say that at least one of them must be false. Similarly, $\sim (P \text{ or } Q) \equiv (\sim P)\&(\sim Q)$. {How would you express it with words?}

A.1.4. To prove $P \Rightarrow Q$, it is the same to prove $\sim Q \Rightarrow \sim P$ (called the **contrapositive form** of $P \Rightarrow Q$). {For example, to prove that, for integers x, y,

$$x \geq 0 \,\&\, y \geq 0 \;\Rightarrow\; xy \geq 0,$$

it is the same to prove that $xy < 0 \;\Rightarrow\; x < 0$ or $y < 0$.}

A.1.5. A proof of $P \Rightarrow Q$ by **contradiction** proceeds as follows. We are given that P is true and we are to show that Q is true. We *assume to the contrary* that Q is false. We now have *two* hypotheses: P and $\sim Q$ (or *one* 'compound' hypothesis $P \,\&\, (\sim Q)$). Using authorized logical maneuvers (like those above), we produce a proposition R such that both R and $\sim R$ are true, which is absurd. The premise $P \,\&\, (\sim Q)$ is therefore untenable; in other words, if P is true then $\sim Q$ must be false (i.e., Q must be true), which is what we wanted to show.

Proofs involving the contrapositive form occur frequently; proofs by contradiction are relatively rare (many proofs containing the phrase "assume to the contrary" turn out, on inspection, to be proofs in contrapositive form).

A.1.6. Finally, we mention proofs by induction. For each positive integer n, we are given a proposition P_n. We wish to show that P_n is true for every positive integer n. This is accomplished in two steps:

(1) verify that P_1 is true;

(2) verify that *if* P_k is true *then* P_{k+1} is true (i.e., verify the *implication* $P_k \Rightarrow P_{k+1}$).

Step (2) is called the **induction step**; the hypothesis P_k of the induction step is called the **induction hypothesis**.

For example, let P_n be the proposition asserting that the sum of the first n positive integers is given by the formula $1 + \ldots + n = \frac{1}{2}n(n+1)$. The truth of P_1 is established by the equality $1 = \frac{1}{2} \cdot 1 \cdot 2$. In the induction step, we *assume* that $1 + \ldots + k = \frac{1}{2}k(k+1)$ (the induction hypothesis) and, on the basis of this assumption, we show (by elementary algebra) that $1 + \ldots + k + (k+1) = \frac{1}{2}(k+1)(k+2)$.

A.2. Set Notations

A.2.1. The expression $x \in A$ means that x is an **element** of the **set** A; it is read "x belongs to A". Its negation, written $x \notin A$, means that x is *not* an element of A.

A.2.2. $\{a, b, c, \ldots\}$ denotes the set whose elements are $a, b, c, \ldots$. In particular, $\{a\}$ is the set whose only element is a; such sets are called **singletons**.

A.2.3. For sets A and B, $A \subset B$ means that every element of A is an element of B, that is,

$$x \in A \Rightarrow x \in B;$$

we then say that A is **contained** in B (or that A is a **subset** of B). The relation $A \subset B$ is also written $B \supset A$ (read "B **contains** A", or "B is a **superset** of A"). These are four ways of saying the same thing: $x \in A \Rightarrow x \in B$.

A.2.4. For sets A and B, $A = B$ means that $A \subset B \& B \subset A$, that is,

$$x \in A \Leftrightarrow x \in B$$

(the elements of A are precisely the elements of B).

A.2.5. $\{n \in \mathbb{Z} : -2 \leq n < 5\}$ denotes the set of all integers n such that $-2 \leq n < 5$; explicitly, it is the subset $\{-2, -1, 0, 1, 2, 3, 4\}$ of $\mathbb{Z}$. More generally, if X is a set and if, for each $x \in X$, we are given a proposition $P(x)$, then

$$\{x \in X : P(x)\}$$

denotes the set of all elements x of X for which the proposition $P(x)$ is **true**. This is a common way of forming subsets of a set X (examples are given below). When there is a prior understanding as to the "universal set" X from which the elements are drawn, we can omit it from the notation and write simply $\{x : P(x)\}$. For example, if it is understood that we are talking about real numbers (not just integers), then $\{x : -2 \le x < 5\}$ is an interval of the real line.

A.2.6. $\varnothing$ denotes the **empty set**, that is, the set with no elements (i.e., the set for which the statement $x \in \varnothing$ is always false).

A.2.7. Let A and B be subsets of a set X. The **union** of A and B is the set

$$A \cup B = \{x \in X : x \in A \ \text{ or } \ x \in B\}.$$

The **intersection** of A and B is the set

$$A \cap B = \{x \in X : x \in A \ \& \ x \in B\}.$$

The **complement** of A in X is the set

$$A' = \{x \in X : x \notin A\},$$

also written $\complement A$ or $X - A$. The **difference** 'A minus B' is the set

$$A - B = \{x \in X : x \in A \ \& \ x \notin B\};$$

thus $A - B = A \cap B'$.

Some properties of these notations:

(1) $A \cup B = B \cup A$, $(A \cup B) \cup C = A \cup (B \cup C)$
(2) $A \cap B = B \cap A$, $(A \cap B) \cap C = A \cap (B \cap C)$
(3) $A \cap (B \cup C) = (A \cap B) \cup (A \cap C)$
(4) $A \cup (B \cap C) = (A \cup B) \cap (A \cup C)$
(5) $A \cup A' = X$, $A \cap A' = \varnothing$, $(A')' = A$
(6) $A \subset B \Leftrightarrow A' \supset B'$
(7) $(A \cup B)' = A' \cap B'$, $(A \cap B)' = A' \cup B'$
(8) $A \subset B \Leftrightarrow A \cap B = A \Leftrightarrow A \cup B = B \Leftrightarrow A - B = \varnothing$.

A.2.8. If X is a set, then the **power set** of X, written $\mathcal{P}(X)$, is the set whose elements are the subsets A of X; thus

$$A \in \mathcal{P}(X) \Leftrightarrow A \subset X.$$

For example, $\{-2, -1, 0, 1, 2, 3, 4\} \in \mathcal{P}(\mathbb{Z})$. {If X is a finite set with n elements, then $\mathcal{P}(X)$ has 2^n elements (in forming a subset of X there are two choices for each point: include it or exclude it); whence the term 'power set'.}

A.2.9. If $\mathcal{S}$ is a set of subsets of a set X—in other words, if $\mathcal{S} \subset \mathcal{P}(X)$—the **intersection** of $\mathcal{S}$ is the set of elements common to the sets in $\mathcal{S}$, written

$$\bigcap \mathcal{S} = \{x \in X : x \in A \ \text{ for all } \ A \in \mathcal{S}\};$$

the **union** of $\mathcal{S}$ is the set of elements of X that belong to some set in $\mathcal{S}$, written

$$\bigcup \mathcal{S} = \{x \in X : \ x \in A \ \text{ for some } \ A \in \mathcal{S}\}.$$

A.2.10. For sets X and Y, the **cartesian product** of X and Y (in that order), written X × Y (verbalized 'X cross Y'), is the set of all ordered pairs (x, y) with $x \in X$ and $y \in Y$; concisely,

$$X \times Y = \{(x, y) : \ x \in X, \ y \in Y\}.$$

Equality of ordered pairs means 'coordinatewise equality':

$$(x, y) = (x', y') \ \Leftrightarrow \ x = x' \ \& \ y = y'.$$

For example, if X = $\{1, 2\}$ and Y = $\{2, 4, 5\}$ then

$$X \times Y = \{(1, 2), (1, 4), (1, 5), (2, 2), (2, 4), (2, 5)\}.$$

If X and Y are finite sets with m and n elements, respectively, then X × Y has mn elements.

A.3. Functions

A.3.1. If X and Y are sets, a **function** from X to Y is a rule f that assigns to each element x of X a single element of Y, called the **value** of f at x and denoted $f(x)$.

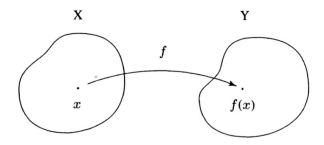

This is expressed compactly by the notation $f : X \to Y$. The set X is called the **initial set** (or **domain**) of f, and Y is called the **final set** of f. The term **mapping** is a synonym for "function" (as are, depending on the context, the terms "transformation" and "operator").

The effect of a function $f : X \to Y$ can also be expressed by

$$f : x \mapsto f(x) \quad (x \in X)$$

(f **sends** x to $f(x)$), or

$$x \mapsto f(x) \quad (x \in X)$$

(x **goes to** $f(x)$). When f is given by a formula, we can even dispense with the letter f; for example, we can speak of the function $\mathbb{Z} \to \mathbb{Z}$ defined by $n \mapsto n^2$.

A.3.2. For any set X the function $X \to X$ defined by $x \mapsto x$ is called the **identity function** on X, denoted $\mathrm{id}_X : X \to X$. If A is a subset of X, the function $A \to X$ defined by $x \mapsto x$ is called the **insertion mapping** of A into X, written $i_A : A \to X$.

A.3.3. If $f : X \to Y$ and $g : Y \to Z$ are functions (g picks up where f leaves off) the **composite** function $g \circ f : X \to Z$ is defined by the formula

$$(g \circ f)(x) = g\big(f(x)\big) \quad \text{for all } x \in X;$$

so to speak, the points of X make the trip to Z by 'transferring' at Y:

$$x \mapsto f(x) \mapsto g\big(f(x)\big).$$

A.3.4. If $f : X \to Y$ is a function and A is a subset of X, the function $A \to Y$ defined by

$$a \mapsto f(a) \quad (a \in A)$$

is called the **restriction** of f to A, denoted

$$f|A : A \to Y.$$

Note that $f|A = f \circ i_A$, where $i_A : A \to X$ is the insertion mapping (A.3.2, A.3.3).

A.3.5. The set of all values of a function $f : X \to Y$ is a subset of Y, namely

$$\{f(x) : \ x \in X\} = \{y \in Y : \ y = f(x) \ \text{for some} \ x \in X\};$$

it is called the **range** (or **image**) of f. For example, the range of the function $f : \mathbb{Z} \to \mathbb{Z}$ defined by $f(n) = |n|$ is the set $\mathbb{N}$ of nonnegative integers.

A.3.6. A function $f : X \to Y$ is said to be **surjective** (or to be a mapping of X **onto** Y) if every element of Y is the image of at least one element of X—so to speak, the function 'uses up' all of the final set. This is merely a matter of notation: the function $\mathbb{Z} \to \mathbb{Z}$ of A.3.5 is not surjective, but the function $\mathbb{Z} \to \mathbb{N}$ defined by the same formula *is* surjective. In general, a function $f : X \to Y$ can be converted into a surjective function by replacing Y by the range of f (so to speak, by throwing away the irrelevant part of the final set). This is possible *in theory* but that doesn't mean we can always *do* it. {What is the range of the polynomial function $p : \mathbb{R} \to \mathbb{R}$ defined by the formula $p(x) = x^6 - 3x^5 + 2x^4 + 5x - 7$?}

A.3.7. A function $f : X \to Y$ is said to be **injective** (or to be a **one-one** mapping) if different elements of X get sent to different elements of Y, that is,

$$x \neq x' \;\Rightarrow\; f(x) \neq f(x') \,;$$

expressed in contrapositive form,

$$f(x) = f(x') \;\Rightarrow\; x = x' \,.$$

If f is not injective, it can't be made injective by a cosmetic change (as in A.3.6); some surgery on the domain is required. For example, the function $\mathbb{Z} \to \mathbb{Z}$ defined by $n \mapsto n^2$ is not injective, but the function $\mathbb{N} \to \mathbb{Z}$ defined by the same formula *is* injective. {A fancier example: The function $\mathbb{R} \to \mathbb{R}$ defined by $x \mapsto \sin x$ is not injective, but the function $[-\pi/2, \pi/2] \to \mathbb{R}$ defined by the same formula *is* injective.}

A.3.8. A function $f : X \to Y$ is said to be **bijective** (or to be a **one-to-one correspondence**) if it is *both* injective and surjective. For example, a bijective mapping $\{1, 2, 3\} \to \{7, 8, 9\}$ is defined by the assignments $1 \mapsto 8$, $2 \mapsto 7$, $3 \mapsto 9$. {Another example: The function $[-\pi/2, \pi/2] \to [-1, 1]$ defined by $x \mapsto \sin x$ is bijective.}

A.3.9. An injective (surjective, bijective) function is called an **injection** (**surjection**, **bijection**).

A.3.10. If $f : X \to Y$ is a bijection, then for each $y \in Y$ there exists a *unique* $x \in X$ such that $y = f(x)$ (existence by surjectivity, uniqueness by injectivity). The assignment $y \mapsto x$ produces a function $Y \to X$, called the **inverse** of f and denoted

$$f^{-1} : Y \to X \,.$$

If $y = f(x)$ then $f^{-1}(y) = x$. The formulas

$$f^{-1}\big(f(x)\big) = x \,, \quad f\big(f^{-1}(y)\big) = y \qquad (x \in X, \; y \in Y)$$

explain the relation between f and f^{-1}; each undoes what the other does. In the notations of A.3.2 and A.3.3,

$$f^{-1} \circ f = \mathrm{id}_X \,, \quad f \circ f^{-1} = \mathrm{id}_Y \,.$$

{Example: For the bijection $[-\pi/2, \pi/2] \to [-1, 1]$ defined by $x \mapsto \sin x$, the inverse function $[-1, 1] \to [-\pi/2, \pi/2]$ is called the Arcsine function.}
   Note that f^{-1} is also bijective and $(f^{-1})^{-1} = f$.

A.4. Integers

The point of departure in this section (a supplement to Chapter 1) is the definition of the field $\mathbb{R}$ of real numbers (1.4.2); the objective is to give an unambiguous definition of the set of 'positive integers' and honest proofs of its key properties (cf. the remarks in 1.4.4).

The 'row of dominoes' image of the set of positive integers is as follows: the set contains 1 (a first domino); the set contains $n + 1$ whenever it contains n (pushing over a domino causes the next one—there *is* a next one—to go over); the set contains nothing else (pushing over 1 causes them all to go down). The following definition embodies in formal language this idea of a smallest possible set that contains 1 and is closed under addition of 1:

A.4.1. *Definition*. There are sets $S \subset \mathbb{R}$ such that (a) $1 \in S$, and (b) $x \in S \Rightarrow x + 1 \in S$ (for example, $\mathbb{R}$ itself has these properties). Let $\mathcal{S}$ be the set of all such subsets S of $\mathbb{R}$, and define $\mathbb{P}$ to be the intersection of all the sets in $\mathcal{S}$:

$$\mathbb{P} = \bigcap \mathcal{S} = \{x \in \mathbb{R} : x \in S \text{ for all } S \in \mathcal{S}\}.$$

The elements of $\mathbb{P}$ are called **positive integers**; if $n \in \mathbb{P}$ we write $n' = n + 1$ and call n' the *successor* of n. The reader who has seen *Peano's axioms* for the positive integers will recognize them in the following theorem:

A.4.2. Theorem. *The set $\mathbb{P}$ of positive integers has the following properties*:
   (i) $1 \in \mathbb{P}$;
   (ii) $n \in \mathbb{P} \Rightarrow n' \in \mathbb{P}$;
   (iii) *if* $1 \in S \subset \mathbb{P}$ *and* $n \in S \Rightarrow n' \in S$, *then* $S = \mathbb{P}$ (Principle of mathematical induction);
   (iv) *if* $m, n \in \mathbb{P}$ *and* $m' = n'$, *then* $m = n$;
   (v) *for all* $n \in \mathbb{P}$, $1 \neq n'$.

Proof. (i), (ii) Every set in $\mathcal{S}$ (notations as in the preceding definition) has these properties, therefore so does their intersection.
   (iii) The assumption is that $S \subset \mathbb{P}$ and $S \in \mathcal{S}$; an intersection is contained in every intersectee, so $\mathbb{P} \subset S$.
   (iv) If $m' = n'$, that is, $m + 1 = n + 1$, then (adding -1 to both sides) $m = n$.

(v) We first observe that $n \geq 1$ for every $n \in \mathbb{P}$. For, if $S = \{n \in \mathbb{P} : n \geq 1\}$, it is clear that $1 \in S$ and that $n \in S \Rightarrow n' \in S$, therefore $S = \mathbb{P}$ by (iii).

In particular, for every $n \in \mathbb{P}$ we have $n \geq 1 > 0$, therefore $n' - 1 = n > 0$, thus $n' \neq 1$. $\Diamond$

From the proof of the theorem we extract the following useful fact:

A.4.3. Theorem. *For every $n \in \mathbb{P}$ we have $n \geq 1$.*

Property (v) says that 1 is *not* the successor of a positive integer; every other positive integer *is*:

A.4.4. Theorem. *If $k \in \mathbb{P}$ and $k \neq 1$ then $k = n'$ for some $n \in \mathbb{P}$.*

Proof. Let $S = \{1\} \cup \{n' : n \in \mathbb{P}\}$. We have $1 \in S$ and it is clear that $n \in S \Rightarrow n' \in S$, therefore $S = \mathbb{P}$ by the principle of induction (property (iii) of A.4.2). In other words,

$$\mathbb{P} = \{1\} \cup \{n' : n \in \mathbb{P}\},$$

whence the assertion of the theorem. $\Diamond$

The next two theorems establish that $\mathbb{P}$ is closed under addition and multiplication (not at all obvious from the definition of $\mathbb{P}$!).

A.4.5. Theorem. *If $m, n \in \mathbb{P}$ then $m + n \in \mathbb{P}$.*

Proof. We have $m + 1 = m' \in \mathbb{P}$ for all $m \in \mathbb{P}$. Let

$$S = \{n \in \mathbb{P} : m + n \in \mathbb{P} \text{ for all } m \in \mathbb{P}\}.$$

By the preceding remark, $1 \in S$. If $n \in S$ then, for every $m \in \mathbb{P}$,

$$m + n' = m + n + 1 = m' + n \in \mathbb{P}$$

(because $m' \in \mathbb{P}$ and $n \in S$), therefore $n' \in S$. By the principle of induction, $S = \mathbb{P}$, whence the assertion of the theorem. $\Diamond$

A.4.6. Theorem. *If $m, n \in \mathbb{P}$ then $mn \in \mathbb{P}$.*

Proof. Let
$$S = \{n \in \mathbb{P} : mn \in \mathbb{P} \text{ for all } m \in \mathbb{P}\}.$$
Obviously $1 \in S$. If $n \in S$ then, for every $m \in \mathbb{P}$,

$$mn' = m(n + 1) = mn + m,$$

where $mn \in \mathbb{P}$ (because $n \in S$), therefore $mn' \in \mathbb{P}$ by A.4.5. Thus $S = \mathbb{P}$ and the theorem is proved. $\Diamond$

A.4.7. *Definition.* We write $\mathbb{Z} = \{m - n : m, n \in \mathbb{P}\}$ and call the elements of $\mathbb{Z}$ **integers**.

In particular, $0 = 1 - 1 \in \mathbb{Z}$; for all $n \in \mathbb{P}$, $n = n' - 1 \in \mathbb{Z}$, thus $\mathbb{P} \subset \mathbb{Z}$; from $-(m-n) = n - m$ we see that $\mathbb{Z}$ 'contains negatives'; and it is easy to see from A.4.5 and A.4.6 that $\mathbb{Z}$ is closed under addition and multiplication. The positive integers (i.e., the elements of $\mathbb{P}$) are precisely the integers (i.e., elements of $\mathbb{Z}$) that are positive for the order of $\mathbb{R}$:

A.4.8. Theorem. $\mathbb{P} = \{x \in \mathbb{Z} : x > 0\}$.

Proof. If $n \in \mathbb{P}$ then $n \in \mathbb{Z}$ and $n > 0$ (A.4.3); thus $\mathbb{P}$ is contained in the set on the right. To prove the reverse inclusion, consider the set

$$S = \{n \in \mathbb{P} : (m \in \mathbb{P} \ \& \ m - n > 0) \ \Rightarrow \ m - n \in \mathbb{P}\};$$

we have to show that $S = \mathbb{P}$.

If $m \in \mathbb{P}$ and $m - 1 = x > 0$, then $m = x + 1 > 1$; by A.4.4, $m = k'$ for some $k \in \mathbb{P}$, thus $x + 1 = m = k + 1$, whence $x = k \in \mathbb{P}$, that is, $m - 1 \in \mathbb{P}$. This proves that $1 \in S$.

Suppose $n \in S$. If $m \in \mathbb{P}$ and $m - n' = x > 0$, then

$$m = n' + x > n' > n \geq 1,$$

so $m = k'$ for some $k \in \mathbb{P}$; then

$$k - n = k' - n' = m - n' = x > 0,$$

therefore $k - n \in \mathbb{P}$ (because $n \in S$), in other words, $m - n' \in \mathbb{P}$. This shows that $n \in S \ \Rightarrow \ n' \in S$, whence $S = \mathbb{P}$. ◊

At this point we call on the completeness axiom:

A.4.9. Theorem. $\mathbb{P}$ *is not bounded above in* $\mathbb{R}$.

Proof. Assume to the contrary that $\mathbb{P}$ is bounded above and let $M = \sup \mathbb{P}$ (1.4.1). In particular, $n + 1 \leq M$ for all $n \in \mathbb{P}$, so $M - 1$ is also an upper bound for $\mathbb{P}$, whence $M \leq M - 1$, leading to the absurdity $1 \leq 0$. ◊

A.4.10. Theorem. (Well-ordering property) *If* S *is a nonempty subset of* $\mathbb{P}$, *then* S *has a smallest element.*

Proof. Assume to the contrary that S has no smallest element. Let N be any element of S; a contradiction to A.4.9 will be obtained by showing that $k < N$ for every $k \in \mathbb{P}$.

It will suffice to show that for every $k \in \mathbb{P}$, there exists an $n \in S$ such that $N - n \geq k$ (for then $N \geq n + k > k$). Let

$$T = \{k \in \mathbb{P} : N - n \geq k \text{ for some } n \in S\};$$

the claim is that $T = \mathbb{P}$ and the proof is by induction.

By assumption S has no smallest element, so there is an $n \in S$ with $n < N$, therefore $N - n \in \mathbb{P}$ (A.4.8) and so $N - n \geq 1$ (A.4.3). This shows that $1 \in T$.

Suppose $k \in T$. Choose $n \in S$ with $N - n \geq k$. By assumption, there is an $m \in S$ with $m < n$, whence $n - m \in \mathbb{P}$, $n - m \geq 1$; then

$$N - m = (N - n) + (n - m) \geq k + 1 = k',$$

which shows that $k' \in T$. This completes the proof that $T = \mathbb{P}$ and achieves the contradiction. $\Diamond$

A.4.11. *Remarks.* **Peano's axioms** for the positive integers[1] postulate the existence of a set $\mathbb{P}$ with an element 1 and a mapping $n \mapsto n'$ satisfying the conditions (i)–(v) of A.4.2. One then has the (arduous) task of defining addition and multiplication and proving that they have the desired properties (associativity, commutativity, distributive law, cancellation, etc.).[2]

In effect, what we have shown in this section is that if one postulates the existence of a complete ordered field $\mathbb{R}$ (a rather high-handed thing to do, actually ...) then the integers come free of charge. This is not news (the reals can be constructed from the positive integers[2]) but it is a welcome simplification as long as the reals are going to be taken axiomatically anyway.

---

[1] Giuseppe Peano (1858-1932).
[2] Cf. E. Landau, *Foundations of analysis* [Chelsea, New York, 1951].

Index of Notations

Index

Printed in the United Kingdom
by Lightning Source UK Ltd.
110557UKS00005B/116

9 780387 942179

CL

515
BER